T0338428

Introduction to Astronomical Spectroscopy

Spectroscopy is the principal tool used in astronomy to investigate the Universe beyond Earth's atmosphere. Through the analysis of electromagnetic radiation, spectrographs enable observers to assess the chemical composition, kinematics, and local physical properties of distant stars, nebulae, and galaxies. Thoroughly illustrated and clearly written, this handbook offers a practical and comprehensive guide to the different spectroscopic methods used in all branches of astronomy, at all wavelengths from radio to gamma-ray, and from ground and space-borne instruments. After a historical overview of the field, the central chapters navigate the various types of hardware used in spectroscopy. In-depth descriptions of modern techniques and their benefits and drawbacks help you choose the most promising observation strategy. The handbook finishes by assessing new technologies and future prospects for deep-sky observation. This text is an ideal reference for today's graduate students and active researchers, as well as those designing or operating spectroscopic instruments.

IMMO APPENZELLER is Emeritus Professor of Astronomy at the University of Heidelberg, Germany, and Director of the Heidelberg State Observatory.

Cambridge Observing Handbooks for Research Astronomy

Today's professional astronomers must be able to adapt to use telescopes and interpret data at all wavelengths. This series is designed to provide them with a collection of concise, self-contained handbooks, which covers the basic principles peculiar to observing in a particular spectra region, or to using a special technique or type of instrument. The books can be used as an introduction to the subject and as a handy reference for use at the telescope or in the office.

Series editors
Professor Richard Ellis, Department of Astronomy, *California Institute of Technology*
Professor Steve Kahn, Department of Physics, *Stanford University*
Professor George Rieke, Steward Observatory, *University of Arizona, Tuscon*
Dr Peter B. Stetson, Herzberg Institute of Astrophysics, *Dominion Astrophysical Observatory*, Victoria, British Columbia

Books currently available in this series:

1. *Handbook of Infrared Astronomy*
   I. S. Glass

4. *Handbook of Pulsar Astronomy*
   D. R. Lorimer and M. Kramer

5. *Handbook of CCD Astronomy*, Second Edition
   Steve B. Howell

6. *Introduction to Astronomical Photometry*, Second Edition
   Edwin Budding and Osman Demircan

7. *Handbook of X-ray Astronomy*
   Edited by Keith Arnaud, Randall Smith, and Aneta Siemiginowska

8. *Practical Statistics for Astronomers*, Second Edition
   J. V. Wall and C. R. Jenkins

9. *Introduction to Astronomical Spectroscopy*
   Immo Appenzeller

Introduction to Astronomical Spectroscopy

IMMO APPENZELLER

Center for Astronomy
University of Heidelberg

CAMBRIDGE
UNIVERSITY PRESS

CAMBRIDGE
UNIVERSITY PRESS

University Printing House, Cambridge CB2 8BS, United Kingdom

One Liberty Plaza, 20th Floor, New York, NY 10006, USA

477 Williamstown Road, Port Melbourne, VIC 3207, Australia

314-321, 3rd Floor, Plot 3, Splendor Forum, Jasola District Centre, New Delhi - 110025, India

103 Penang Road, #05-06/07, Visioncrest Commercial, Singapore 238467

Cambridge University Press is part of the University of Cambridge.

It furthers the University's mission by disseminating knowledge in the pursuit of education, learning and research at the highest international levels of excellence.

www.cambridge.org
Information on this title: www.cambridge.org/9781107015791

© Immo Appenzeller 2013

First published 2013

A catalogue record for this publication is available from the British Library

Library of Congress Cataloging in Publication data
Appenzeller, I. (Immo), 1940–
Introduction to astronomical spectroscopy / Immo Appenzeller.
pages   cm. – (Cambridge observing handbooks for research astronomers ; 9)
Includes bibliographical references and index.
ISBN 978-1-107-01579-1 (hardback) – ISBN 978-1-107-60179-6 (paperback)
1. Astronomical spectroscopy.   I. Title.
QB465.A67   2013
522′.67–dc23   2012019846

ISBN 978-1-107-01579-1 Hardback
ISBN 978-1-107-60179-6 Paperback

Contents

Preface

With the exception of a few objects that have been successfully identified as sources of highly energetic charged particles or of neutrinos, all our knowledge about the universe outside the inner solar system is based on the analysis of electromagnetic radiation. Some valuable information has been derived by measuring the flux, the time variations, or the polarization of astronomical radiation sources. By far the most important tool for investigating cosmic objects, however, has been the analysis of their energy distributions and of their line spectra. There are obvious reasons for this predominance of spectroscopic methods in modern astronomy. First, spectra contain a particularly large amount of physical information. If properly analyzed, spectra allow us to determine the chemical composition, local physical conditions, kinematics, and presence and strength of local physical fields. Second, apart from the cosmological redshift and the reduced observed total flux of faraway objects, spectra are independent of the distance, making spectroscopy a particularly valuable remote-sensing tool. Finally, there exists a well-developed theory of the formation of continua and line spectra.

The gathering of information on distant objects by means of spectral observations requires several steps. First, suitable instruments must be designed that allow us to measure the spectra of the faint astronomical sources. Then, these instruments must be employed to obtain spectra of optimal quality. Finally, the spectra must be analyzed and physical information on the observed objects must be extracted. In astronomy, the term *spectroscopy* is used for all these steps. Moreover, the theory of radiative transfer and line formation in stellar atmospheres is often included in the term *astronomical spectroscopy*. The content of this book is focused on the first two of these topics – that is, on the techniques and the practice of obtaining spectra of cosmic objects. Apart from space restrictions, there are several reasons for this emphasis on observational aspects. One obvious reason is the fact that this book belongs to a series of *observing handbooks*. Moreover, whereas astronomers observing different types of objects often use the same or similar instrumentation, the methods used for analyzing the spectra usually depend on the nature of the radiation sources. Because modern astronomy comprises many different types of objects, including stars, black holes, interstellar gas and dust, galaxies, and a variety of nonthermal emitters, the methods for analyzing astronomical spectra are correspondingly diversified. It would be difficult to treat them adequately in a single volume. Finally, there already are excellent books describing the analysis of astronomical spectra of different types. Examples of these books are cited in the second chapter of this volume.

Astronomical spectroscopy started in the nineteenth century in the optical spectral range. It later spread to all wavelengths at which astronomical observations are possible. In the early days, the observers typically built their own instruments and were specialists for a single wavelength regime. Today, much of the cutting-edge science is done at large ground-based and space-based observatories that are accessible to large observer communities, and present-day observers typically carry out observations at many different wavelengths. Therefore, this book tries to include all wavelengths at which astronomical spectroscopy is possible at present. It aims at providing all information that a researcher needs to plan and to execute astronomical spectroscopic observations with existing instrumentation. It provides the level of technical detail that is required for selecting the techniques and methods that are best suited for a given task. By describing the present-day technical state of the art, along with the limitations of present instruments, the book can furthermore be a good introduction to the topic for scientists and engineers who aim at improving present methods and on developing future, superior instrumentation. Finally, the book will, hopefully, help the reader to better understand and appreciate the opportunities of the vast amount of spectroscopic data that are available in astronomical data archives.

As this series is written for research astronomers, it is expected that the reader has some basic knowledge of astronomy and of elementary physics. Moreover, it will be assumed that the reader is familiar with the special notations and definitions that are frequently used in astronomy. Some of these notations and definitions have been changing with time. This book normally follows the notations currently defined by the International Astronomical Union (IAU) and specified on the IAU Web sites. For historic reasons, some technical terms are used with a different meaning in the astronomical literature. In these cases, I adopted the meaning that is preferred by a majority of the professional astronomers in the current scientific literature. Finally, topics that have been discussed in other volumes of this series of observing handbooks will not be treated here in detail, but references will be given in all such cases.

Present-day science depends on cooperation and teamwork. This book is no exception. It could not have been completed without the kind support of many different colleagues. Special thanks are due to Joachim Krautter, Walter Seifert, and Otmar Stahl, for critically reading parts of the manuscript. Moreover, I would like to thank Regina von Berlepsch, Luis Carrasco, Frank Eisenhauer, Edith Falgarone, Kay Justtanont, Norbert Kappelmann, Andreas Kaufer, Richard Kron, Olivier Le Fèvre, Holger Mandel, Michel Mayor, Karl Menten, Patrick Osmer, Jürgen Schmitt, Walter Seifert, Josef Solf, Otmar Stahl, Matthias Steinmetz, Stefan Wagner, Wenli Xu, and the Astrium GmbH for providing figures and/or for allowing me to use their figures in this book. Thanks are also due to the editors of *Astronomy and Astrophysics* (A & A) and to ESA, ESO, NASA, Fermilab, the H.E.S.S. consortium, and the Gemini Observatory for the kind permission to reprint illustrations.

1

Historical Remarks

The purpose of this book is to provide an introduction to present-day astronomical spectroscopy. Thus, this chapter on the historical development will be restricted to a brief outline of selected milestones that provided the basis for the contemporary techniques and that are helpful for an understanding of the present terminologies and conventions. The reader interested in more details of the historic evolution of astronomical spectroscopy may find an extensive treatment of this topic in two excellent books by John Hearnshaw (1986, 2009). Additional information can be found in older standard works on astronomical spectroscopy, which were published by Hiltner (1964), Carleton (1976), and Meeks (1976). Apart from (still up-to-date) historical sections, these books provide extensive descriptions of methods that have been used in the past, before they were replaced by the more efficient contemporary techniques.

1.1 Early Pioneers

Astronomy is known for its long history. Accurate quantitative measurements of stellar positions and motions were already carried out millennia ago. On the other hand, spectroscopy is a relatively new scientific tool. It became important for astronomical research only during the past 200 years. The late discovery of spectroscopy may have been due to the scarcity of natural phenomena in which light is decomposed into its different colors. Moreover, for a long time the known natural spectral effects were not (or not correctly) understood. A prominent example is the rainbow. Reports of rainbows and thoughts about their origin are found in the oldest known written texts, and in most parts of the world almost everybody alive has seen this phenomenon. That rainbows are somehow caused by the reflection of sunlight in raindrops was suggested already by the ancient Greek and Arabian astronomers, and the basic geometry of the rainbow

1

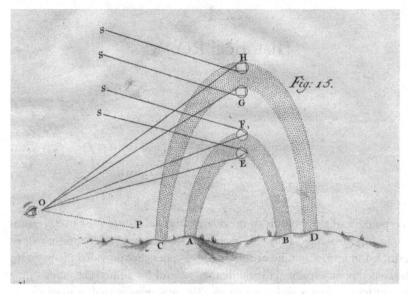

Figure 1.1. Newton's explanation of the colors of the primary and secondary rainbow as spectra of the sunlight produced by refraction and internal reflection in spherical raindrops. From Newton (1740).

(shown in Figure 1.1, taken from Newton's work) was described correctly already at the beginning of the fourteenth century by the Dominican friar Theodoric of Freiberg (about 1250–1310; see, e.g., Grant, 1974), who concluded that rainbows result from a combination of reflection and refraction of sunlight in small spheres of water. The correct explanation of the observed colors as a solar spectrum, however, was given only about three centuries ago by Isaac Newton (1642–1722).

Newton based his conclusions on extensive studies of the refraction of light in lenses and prisms. For most of his experiments, Newton used the light of the Sun. Hence, Newton not only started spectroscopy, but he also was the first one to study the spectrum of a celestial object.

Newton described his optical experiments and their results in the large three-volume work *Opticks*, which first appeared in 1704, followed by several improved and corrected editions. Later, *Opticks* was translated into Latin, which in Newton's time still was the *lingua franca* of science. The Latin version, titled *Optice*, became widely known outside England. Figures 1.1 and 1.2 have been taken from this Latin edition.

Figure 1.2 shows some of the early optical arrangements that Newton used to obtain solar spectra. In his first experiments, Newton produced a circular

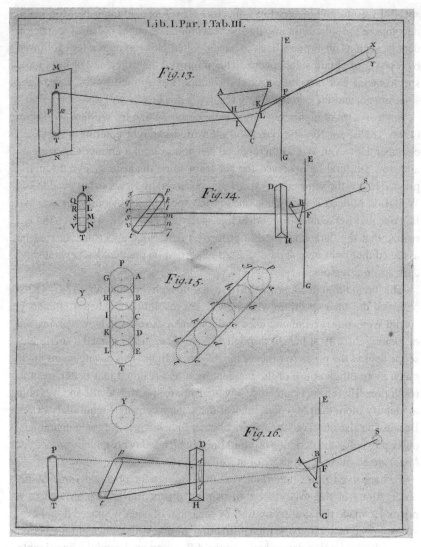

Figure 1.2. Optical arrangements used by Isaac Newton to produce spectra of the Sun, as published and described in *Optice* (Newton, 1740).

spot of sunlight on the wall of his experiment chamber by means of a hole in the sunlit shutter at the opposite side of the room. He then placed one or several prisms into this quasi-parallel light beam. As illustrated by his drawings (reproduced in Figure 1.2), the result was a solar spectrum. However, the size of the beam (with a diameter of approximately one inch) severely limited the spectral resolution[1] of Newton's spectra. Therefore, although Newton could decompose the solar light into its different colors, he could not resolve spectral details, such as absorption lines or bands. Later, Newton used a lens followed by a prism to obtain solar spectra. This improved the efficiency of his simple spectrometer, but now the spectral resolution was limited by the size of the solar image, and the corresponding resolving power was still low.

Solar spectra of higher resolution were observed first by William Wollaston (1766–1828). As Newton did, Wollaston used sunlight entering a darkened chamber. However, Wollaston replaced Newton's circular entrance aperture by a narrow slit, and he observed the slit from some distance by eye through a prism. He saw the same color sequence as observed by Newton, but in addition, Wollaston noticed several dark lines in the spectrum. Obviously, he was the first one to observe an astronomical line spectrum. However, Wollaston did not realize that the dark lines were a property of the solar spectrum. Instead he assumed that light in general consisted of several distinct colors, separated by dark spectral regions.

The breakthrough came in 1814 when Joseph Fraunhofer (1787–1826) observed the solar spectrum with even higher resolution and better efficiency. As did Wollaston, Fraunhofer used sunlight and a narrow slit. However, instead of looking directly at the slit through a prism, Fraunhofer placed a small tele-scope behind the prism to observe the dispersed light. This arrangement made much more efficient use of the incident light and resulted in a better spectral resolution. Therefore, Fraunhofer saw many more spectral details and was able to identify more than 500 dark lines of the solar spectrum (Fraunhofer, 1817). Because he also observed the spectra of other light sources, he could unambigu-ously prove that the dark (absorption) lines (which today are called *Fraunhofer lines*) are an intrinsic property of the solar spectrum.

Perhaps even more important was that Fraunhofer obtained some of his observations of the solar spectrum using a diffraction grating instead of a prism. He made this grating himself by means of of equally spaced thin wires. As explained in Section 3.3, for a diffraction grating with known grating period there exists a simple relation between the diffraction angle and the wavelength of the diffracted light. As a result, Fraunhofer could directly measure and list the wavelengths of the observed solar absorption lines.

By placing his prism in front of an astronomical telescope, Fraunhofer also succeeded in observing the spectra of several bright stars and of the planet Venus. He found that the spectrum of Venus was identical with the solar spectrum, but that the observed stars showed very different spectra.

---

[1] For an exact definition of this term, see Section 3.3.

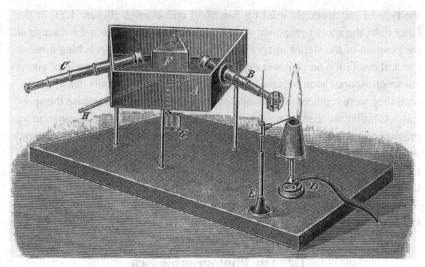

Figure 1.3. The prism spectrometer used in 1859 by G. R. Kirchhoff and R. W. Bunsen to develop spectrum analysis (according to Kayser, 1900). In this arrangement, spectra of various substances were produced by heating corresponding probes in the gas flame D. The light entering the spectrograph slit (at the front end of the tube B) was then collimated to a parallel beam by means of a biconvex lens. The collimated beam was dispersed into its colors by the prism F, and the resulting spectrum was observed visually by means of the small telescope C.

He noted, for instance, that the spectrum of Sirius contained very few lines, which were particularly broad and strong, however. Obviously, by observing Sirius Fraunhofer had, for the first time, seen the broad hydrogen absorption lines that are typical of A-type stellar spectra.

The next important milestone in the development of astronomical spectroscopy took place in 1859 in a chemical laboratory. In that year, the physicist Gustav Kirchhoff (1824–1887) and the chemist Robert Bunsen (1811–1899) invented the method of spectrum analysis. Using the spectrometer reproduced in Figure 1.3, they systematically investigated the line spectra of various chemical elements. Comparing their laboratory results with the solar spectrum, they found that many of the wavelengths and spectral patterns observed in the lab experiments were identical to those reported by Fraunhofer in the solar spectrum. They correctly concluded that the Fraunhofer lines were signatures of the chemical composition of the solar surface layers. Although the technique was initially developed for laboratory measurements, the great potential of spectrum analysis for astronomical research was recognized immediately, and many different astronomers now started observing stellar spectra. They used either Fraunhofer's arrangement (with a large prism in front of a telescope) or the type of spectrometer used by Kirchhoff and Bunsen (Figure 1.3). In the latter case, the spectrograph slit was placed in the focal plane of a telescope at the position of the stellar image. A slit had the advantage of reaching a better spectral resolution and a lower sky background contribution. Therefore, visual slit spectroscopes soon became the standard spectral instruments in astronomy, until they were replaced by photographic spectrographs. To increase the spectral resolution, instead of a single prism often a train of several consecutive prisms was placed in the parallel beam behind the collimator. Among the many astronomers who (following the work of Fraunhofer, Kirchhoff, and Bunsen) used and improved the technique of visual astronomical spectroscopy were G. B. Donati and Angelo Secchi in Italy, Lewis M. Rutherfurd in the United States, George Airy and William Huggins in the United Kingdom, Jules Janssen in France, and Hermann Carl Vogel in Germany.

1.2 The Photographic Era

With the visual spectroscopes, many basic properties of stellar spectra were discovered, and the first spectral classification schemes were developed. However, because the spectra had to be recorded by making drawings and notes at the telescope and by measuring each individual spectral line, visual observations were cumbersome and time consuming. Moreover, only relatively bright stars could be observed by the eye. Therefore, as soon as the new technique of photography was invented in the middle of the nineteenth century, first attempts were made to use photography for recording spectra. Photographic spectra of the Sun were obtained in 1842 by Alexandre-Edmond Becquerel (1820–1898) in France and in 1843 by John W. Draper (1811–1882) in the United States. However, it took another thirty years before John W. Draper's son, Henry Draper (1837–1882), succeeded in obtaining the first usable photographic spectrum of a star in 1872. This success came so late because the early photographic techniques were much less sensitive than the human eye. Even the most sensitive photographic emulsions (developed during the twentieth century) converted only about 1 percent of the incident light into a usable signal. This is about the same fraction as the human eye can detect. However, although their light sensitivity is low, photographic emulsions have the immense advantage of being able to integrate the light over minutes or (depending on the plate properties) even hours. Moreover, in the last quarter of the nineteenth century new types of photographic plates became available, which finally were sufficiently sensitive for recording stellar spectra. As a result, by the end of the nineteenth century photographic plates had become the standard light detectors of astronomy and astronomical spectroscopy.

Figure 1.4. The photographic spectrograph that Hermann Carl Vogel installed in 1888 at the 30-cm refractor of the Astrophysical Observatory at Potsdam. The tube extending from the telescope contains the slit and the collimator optics. The curved box houses the prism train. At the lower left end of the prism box, the photographic camera is attached. The small telescope at right was used for adjustments and guiding. Image courtesy Leibniz Institute of Astrophysics, Potsdam.

After Henry Draper's pioneering work, photographic spectroscopy was developed further mainly by William and Margaret Huggins at their private observatory near London, and later by Hermann Carl Vogel (1841–1907) in Potsdam. In 1873, while working at a small private observatory in northern Germany, Vogel had started the first big (still visual) survey of the spectra of all bright stars. He continued this survey (in cooperation with others) when in 1874 he moved to the newly founded Astrophysical Observatory at Potsdam (AOP), where he became the director in 1882. Having learned of the work of Draper and Huggins, Vogel realized that his spectroscopic survey could be improved and accelerated using photographic spectra. His first photographic observations were made in 1887 with an experimental two-prism spectrograph. In 1888, this instrument was replaced by the spectrograph that is shown in Figure 1.4. This instrument, which was built by professional opticians according to Vogel's design, became the model and prototype of many later photographic spectrometers.

Vogel's instrument (and later photographic spectrographs) still basically followed the optical arrangement of Kirchhoff and Bunsen. However, the small telescope used for observing the spectra visually (C in Figure 1.3) was now replaced by a photographic camera. Moreover, during the twentieth century, for most applications prisms were gradually substituted by diffraction gratings as light-dispersing optical elements.

Relative to prisms, gratings had several important advantages. As explained in Chapter 3, with gratings a much larger range of dispersions can be achieved. Moreover, the dispersion of a grating spectrograph is a linear function of the wavelength, whereas the dispersion of prisms decreases rapidly with increasing wavelength and, for most materials, becomes very small in the red spectral range. Finally, gratings can be used at any wavelength between X-rays and the far infrared (FIR), whereas prisms are restricted to the spectral ranges for which transparent prism materials exist.

As noted earlier, a diffraction grating had been used in 1821 by Joseph Fraunhofer to determine the wavelengths of the solar absorption lines. During the second half of the nineteenth century, grating spectroscopes became the standard tools for solar spectroscopy. Perhaps the best-known examples of the early solar grating spectrometers are the instruments of H. A. Rowland (1848–1901), who, as the (first) professor of physics at the Johns Hopkins University in Baltimore, produced the best diffraction gratings of his time and used them to compile an extensive catalog of the absorption lines of the solar spectrum.

Early attempts to use grating spectroscopes for stellar spectroscopy were much less successful. The reason for these failures was the low efficiency of the first gratings. All early gratings were transmission gratings, where most of the light ended up in the undispersed 0th order. Therefore, they were much less efficient than the prisms of this time. This changed dramatically when blazed reflection gratings (see Chapter 3) became available in 1912. During the following decades, prism and grating spectrographs were still used (and developed) in parallel. After 1935, though, almost all new instruments used diffraction gratings.

Another important new element of astronomical spectrographs were the wide-field Schmidt cameras, which became available after 1930. With their large field and their short focal length, Schmidt cameras were particularly useful in combination with photographic plates as detectors. They soon became the standard cameras for small, efficient low-resolution Cassegrain spectrographs, as well as for the huge stationary coudé-focus instruments that were built for the leading (2- to 5-m) telescopes during the first half of the twentieth century.

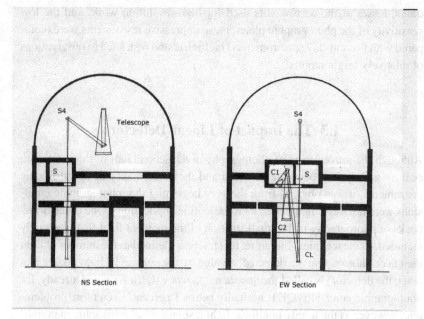

Figure 1.5. Example of a large photographic Coudé spectrograph. The figure shows two cross sections of the building of the 2.2-m telescope of the Calar Alto observatory in southern Spain. The light from the telescope is fed into the spectrograph by means of the stationary mirror S4 on the extension of the (fork-mounted) telescope's polar axis. After passing the spectrograph slit S, the light beam is collimated by the concave mirror CL at the ground floor level of the building, and spectra are recorded by means of a reflection grating (above C1) and one of the two large Schmidt cameras C1 or C2. Note the size of the spectrograph relative to the 2.2-m telescope. Adapted from Bahner and Solf (1972).

An example is the optical layout of the coudé spectrograph of the 2.2-m telescope at the Calar Alto Observatory in Spain, reproduced in Figure 1.5. At the time of its completion, this instrument was one of the most advanced astronomical high-resolution spectrometers. As shown in the figure, such spectrographs sometimes were significantly larger than the telescopes to which they were attached. The large, stationary coudé instruments reached resolving powers of up to 80,000, and the high-quality photographic spectra[2] obtained with these spectrographs formed the basis for the development of the theory of stellar atmospheres. On the other hand, because of light losses in the coudé mirror

---

[2] In the older astronomical literature, the term *spectrum* was used exclusively for the wavelength or frequency distribution of light, whereas the recording of a spectrum was called a *spectrogram*. Following the present usage in the astronomical literature, both items are called *spectrum* in this book.

trains, losses at the narrow slits used for high-resolution work, and the low sensitivity of the photographic plates, these impressive instruments were (compared with present-day spectrometers) inefficient and restricted to observations of relatively bright targets.

1.3 The Impact of Linear Detectors

Although the introduction of photography in the second half of the nineteenth century greatly improved the potential and the performance of spectroscopic instruments, it was obvious from the very beginning that photographic emulsions were not ideal light detectors for scientific applications. One of the drawbacks of photography is the small fraction of the incident light that is actually recorded. In the technical literature, this fraction (defined as the number of light quanta or photons that are detected, divided by the number of light quanta that reach the detector) is called the *quantum efficiency* (QE). As noted already, for photographic emulsions QE is normally below 1 percent. Even more problematic, however, is the highly nonlinear light response of photographic materials. Normal photographic emulsions are completely insensitive to light levels below a certain threshold, independent of the exposure time. At very high light levels, the opaqueness of the developed plate becomes independent of the illumination, or it even decreases again with increasing light intensity ("overexposure" and "solarization" effects). A monotonic relation between the incident light intensity and the opaqueness of the developed and fixed plate exists only in a relatively small intensity range. To make matters worse, this relation depends on the chemicals used, the development time, the age of the plate, the temperature, and various other environmental parameters. To understand these unwelcome properties of photographic plates and the progress that resulted from the new detectors, in the following paragraphs the basic mechanisms of photon detection are briefly summarized in terms of solid-state physics.

1.3.1 Photon Detection

Photography is based on the internal photoeffect in small semiconductor crystals. Its properties can be understood from the energy states of electrons in a semiconductor, which are outlined schematically in Figure 1.6. At zero absolute temperature, all electrons of an ideal semiconductor have energies corresponding to the low valence energy band. At higher temperatures, most electron energies are still within the valence energy band, but some electrons are found

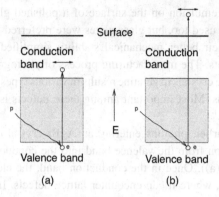

Figure 1.6. Schematic representation of the energy states and the production of conduction electrons or free electrons by the absorption of light in a semiconductor. The electron energy E increases from bottom to top. At low temperatures, almost all electrons of a semiconductor have energies corresponding to the lower (valence) energy band. Only a few electrons are found in the conduction band, which is separated from the valence band by the shaded "forbidden" energy range. Because of quantum effects, electrons with energies in the forbidden range cannot exist inside a solid. The energy difference between the conduction band and the valence band is called the *band gap*. By absorbing a photon with an energy *hν* larger than the band gap, a valence electron can be lifted into the conduction band (Case (a)), or even be ejected from the solid (Case (b)). Case (a) occurs in silver halides during the photographic process and in modern array detectors. Case (b) occurs in photocathodes.

in the conduction band, which is separated from the valence band by a "forbidden" energy range. Because of quantum effects, energies in this forbidden range cannot be attained by electrons inside the semiconductor. Electrons in the conduction band can change their energy. By acquiring kinetic energy, they can move and produce an electric current. The energy difference between the conduction band and the valence band is called the *band gap*.

If a photon with an energy *hν* larger than the band gap enters the semiconductor, it can can transfer its energy to an individual valence band electron. As a result, an electron can be lifted into the conduction band (case (a) in Figure 1.6), or it can even escape from the solid (case (b)). Both processes are called *photoeffect*. Case (a) occurs in the photographic process and in modern solid-state array detectors (such as silicon-based charge-coupled devices [CCDs]) and is called the *internal photoeffect* or *photoconduction effect*. Case (b), which is called the *external photoeffect* or *photoelectric effect*, occurs in photocathodes.

In photographic emulsions, microcrystals of the semiconductor silver bromide (AgBr), or another silver halide, are usually used. These small crystals, with typical sizes of 10–30 micrometers (called "grains"), were suspended in a dried gelatine emulsion on the surface of a polished glass plate. Occasionally films were used too, but glass plates were preferred for astronomical spectroscopy, as their better mechanical stability simplified accurate wavelength measurements. The manufacturing process of photographic plates and the immersion of the crystals in gelatine results in various types of lattice defects in the AgBr crystals. Most important among these defects is the presence of free silver ions (Ag^+).

As explained earlier, photons entering an AgBr crystal can be absorbed by lifting an electron from the valence band into the conduction band of the lattice (Figure 1.6 (a)). Once in the conduction band, the electron can move through the crystal, where it can encounter lattice defects. In particular, if it meets a positive silver ion, a conduction electron can recombine with the ion and form a silver atom. Because of special properties of the AgBr lattice, the silver ions and atoms can (to some extent) also migrate within the grain. If several silver atoms meet, they form a metallic *silver cluster*. Silver atoms that do not become attached to a silver cluster eventually become ionized again and are reintegrated into the lattice. During the *development* and *fixing* of a photographic plate or film, all grains that contain silver clusters above a certain size are chemically converted into metallic silver, whereas all other grains are dissolved and removed. Because the silver grains are opaque and their presence and number density depend on the illumination pattern during the exposure, the concentration of grains that are converted to metallic silver forms a (negative) image of the illumination pattern.

The first step of this process (the absorption of photons and the production of conduction electrons) is essentially linear. However, the following processes are all nonlinear. Grains contribute to the image only if more than one silver atom has been formed in the grain, and only if a sufficient number of these atoms meet. This explains why the light sensitivity of photographic materials is so low and nonlinear, and why below a certain threshold light has no effect at all on photographic emulsions.

From the experience with the photographic process, it became clear that a more sensitive and more linear semiconductor-based detector required a direct recording of the electrons that are lifted from the valence band by the absorption of photons. This is exactly what is done in the present-day astronomical light detectors, which are described in the following sections.

An advantage of photographic plates was that they formed their own data storage. The experience of by now more than one century has shown that – if properly developed and fixed – plates can be stored without significant deterioration for many decades, whereas the digital spectra of modern detectors must be recopied from time to time to ensure their long-term readability. On the

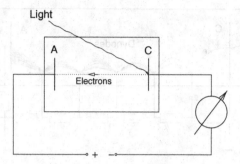

Figure 1.7. The principle of measuring light intensities with photocathode detectors. Light reaching the photocathode (C) produces free electrons at the surface of the cathode. The electrons are attracted by the positive electrode A (the anode) and the resulting electric current is measured. It is directly proportional to the incident light flux.

other hand, digital data have the advantage of being accessible remotely through data networks, whereas plates can be used locally only. Therefore, during the past years many astronomical plates that were taken during the photographic era have been scanned and digitized. Many of these scans are available through the large astronomical data archives and the virtual observatories.

1.3.2 Photocathode Devices

The first linear astronomical detectors made use of photocathodes. Like the silver halide grains of photography, most photocathodes are composed of semiconducting solids. Instead of generating a conduction electron, however, in a photocathode the absorption of a photon results in the ejection of a valence-band electron into the space outside the solid (case (b) in Figure 1.6). Therefore, for photocathode materials the energy that is needed to eject an electron from the solid must be low. In solid-state physics, this energy is called the *work function* of the material. Good photocathodes, obviously, are materials with a small work function. This property of a solid can be achieved by generating (through differential doping) chemical gradients in thin surface layers, which assist the escape of electrons. In some photocathodes there is only a very small difference between the energy of the bottom of the conduction band and the energy at which the electrons can leave the surface of the solid. To prevent the thermal escape of electrons, such materials must be permanently cooled.

The principle of measuring light fluxes with a photocathode is illustrated in Figure 1.7. Light falling onto the photocathode C produces free electrons in the space outside the solid. The negatively charged electrons are attracted by

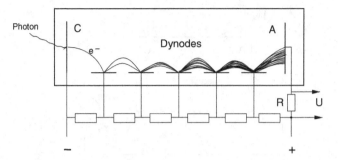

Figure 1.8. The principle of charge amplification in a photomultiplier. Light reaching the transparent photocathode C produces free electrons, which are accelerated toward the first dynode. At the dynode, each incident electron produces several (up to about 50) secondary electrons. At the following dynodes, the secondary electrons generate additional electrons, until a large electron avalanche reaches the anode A. The resulting electric current causes an easily measurable voltage pulse at the resistor R.

the positive electrode A (the anode), where they are absorbed. Each electron contributes one elementary charge to the resulting photocurrent. Therefore, this current is exactly proportional to incident light flux. Because free electrons interact readily with air molecules, the electrodes C and A have must be enclosed in a vacuum tube.

In the laboratory, the photoelectric effect was demonstrated by Heinrich Hertz and his coworkers in 1887. However, it took some time before it became possible to measure the very small electric currents produced by faint astronomical objects. After having been used only occasionally for astronomical photometry during the first half of the twentieth century, photocathode detectors became competitive to photographic plates by about 1940, when *photomultipliers* became available. As indicated by the name, in photomultipliers each primary photoelectron is converted into a large number of secondary electrons. As illustrated by Figure 1.8, for this purpose the primary electrons are accelerated toward additional electrodes (called *dynodes*) with electric potentials between the photocathode and the anode. The potential differences are produced using a chain of resistors between the cathode and the anode. By applying a high voltage that accelerates the electrons to energies many times larger than the work function of the dynode material, at each dynode an incident electron liberates several new electrons. By repeating this process at consecutive dynodes, charge amplifications by typical factors of 10^8 can be achieved. In this way, each primary photo electron produces a short avalanche of secondary electrons with a total charge that is easily measurable. With a sufficiently high time resolution it is then possible to count the individual electron avalanches. Because each charge avalanche corresponds to one primary electron, the number of detected photons can be counted directly, and the photon flux can be measured digitally. Alternatively, the charges of the electron avalanches can be integrated to produce an analog signal that is proportional to the photon flux.

Because of their linearity and peak quantum efficiencies of the order of 30 percent, by the middle of the twentieth century photomultipliers had become the standard detectors for astronomical photometry and for some spectroscopic applications. However, normal photomultipliers are one-pixel detectors without an imaging capability. Therefore, with a conventional photomultiplier, spectra had to be recorded by scanning the spectral range sequentially. Because of the brightness of the Sun, this was acceptable for solar spectroscopy. However, for most stellar work, the less sensitive photographic plates initially remained more efficient, as all wavelengths could be recorded simultaneously.

Sensitive imaging photocathode detectors, in which the charge distribution on the photocathode is electronically imaged onto a position-sensitive anode or onto a scintillation screen (which could be read out by a less sensitive detector), became available by about 1960. For the following two decades, the position-sensitive cathode devices were the standard detectors for astronomical spectroscopy. Although almost forgotten by now, these devices played a fundamentally important role for the progress in astronomy in the second half of the twentieth century. Originally developed for TV cameras, imaging photocathode detectors used a variety of different imaging and readout techniques. Because most of these devices are no longer in use, they will not be discussed here. Readers interested in these historic detectors can find a discussion of their concepts, merits, and problems in an earlier review by this author (Appenzeller, 1989).

For most applications, photocathode devices became obsolete when efficient, large-format solid-state array detectors became available by about 1980. Only for observations at far and extreme ultraviolet (UV) wavelengths and for soft X-rays, photocathode detectors sometimes still have advantages. In principle, silicon and other semiconductors are very sensitive to UV and soft X-ray light. However, the mean free path of UV photons in silicon is so short that the photoeffect takes place close to the surface, where various surface effects can prevent an efficient charge readout. In photocathodes, electron production close to the surface tends to be helpful.

The most frequently used cathode detectors for UV spectroscopy are *microchannel plates* (MCPs), which are imaging detectors based on the principle of photomultipliers. They consist of a large number of tightly spaced parallel tubes with semiconducting inner surfaces. The tube diameters are of the order

Figure 1.9. Front view of a microchannel plate (MCP) detector for FUV spectroscopy. This detector contains a stack of three MCPs with diameters of 6 cm. It
was used in the ORFEUS FUV/EUV spectrometer described in Section 7.2. Image
courtesy Institute of Astronomy, University of Tübingen.

of 10 μm and their maximal length amounts to a few millimeters. Usually
these plates are placed between a semitransparent photocathode and a position-
sensitive anode. By applying suitable voltages to the front and back surfaces of
the plate, photoelectrons from the cathode are drawn into the semiconducting
tubes, where they collide with the tube walls and generate avalanches of secondary electrons. These avalanches are then recorded by the anode as current
pulses. Because the photoelectrons remain confined to the tube that they enter,
the image information from the photocathode is retained during the amplification process. If the channel walls have a photosensitive coating, it is also
possible to produce the primary photoelectrons in the front parts of the channels. In this case, an extra photocathode can be omitted, but the channels must
be slightly inclined relative to the incident light rays to prevent light passing
through the plate without being absorbed.

Compared with a conventional photomultiplier, the wall of each microchannel tube combines the function of the dynode chain, the function of the chain
of resistors that provide the potential gradient for the electron acceleration, and
sometimes the function of the photocathode as well. If the charge amplification
of a single MCP is not sufficient, several consecutive MCPs can be stacked. An
example of an MCP detector, which had been developed for high-resolution far
ultraviolet (FUV) spectroscopy, is shown in Figure 1.9.

Another field in which photocathodes are still in use is high-time-resolution
spectroscopy. Because the array detectors described in the next section require

finite readout times and are affected by readout noise, their efficiency and their signal-to-noise ratio (S/N) tend to become low for very short exposures. Photomultipliers and channel plates are free of readout noise and their time resolution (limited only by the time spread of the electron avalanches) can be much higher than in the case of array detectors.

1.3.3 Solid-State Array Detectors

Solid-state array detectors consist of two-dimensional arrangements of photon-sensitive diodes in semiconductor-based integrated circuits. All these devices make use of the internal photoeffect (Case (a) of Figure 1.6). Thus, in array detectors, conduction-band electrons are produced in the same way as in the grains of photographic emulsions. However, in contrast to the complex "chemical readout" of photography, in array detectors the number of photoelectrons (and the light intensity) is derived directly by recording the resulting electric charge or a resulting electric current. In addition to the photodiodes, the integrated circuits contain the required readout electronics and a signal preamplifier. The best-known, and most important, array detectors are the silicon-based charge-coupled devices (CCDs). However, direct-readout detector arrays (usually referred to as complementary metal oxide semiconductor [CMOS] detectors) are also used widely in astronomy. Whereas at least the readout part of modern CCD detectors is always based on silicon wafers, CMOS detectors can be made of various different semiconductor materials. CCDs and their astronomical applications are described in a dedicated separate volume of this handbook series (see Howell, 2006). Therefore, their technology will not be discussed here. Up-to-date descriptions of the technology of the different types of array detectors, which are in use in present-day astronomy, can be found in the proceedings of the regular workshops on this topic (e.g., Beletic et al., 2006).

A key advantage of array detectors relative to photographic plates and photocathodes is their higher quantum efficiency, which in some devices reaches peak values greater than 95 percent, while the corresponding values are typically less than 30 percent for photocathodes and less than 1 percent for photographic plates. Moreover, using array detectors of different materials, a large wavelength range can be covered. Conventional CCDs, which make use of the photoeffect in silicon, are sensitive to about the same wavelength range as photographic emulsions. However, by combining silicon-based CCD readout circuits with other light-sensitive semiconductors, efficient detectors for the near infrared (NIR) and the mid-infrared (MIR) range can be produced. With CMOS detectors, the range can be extended into the far infrared.

An operational advantage of array detectors is their mechanical robustness and their lower sensitivity to overexposure (relative to photocathodes). As noted earlier, the low work functions of photocathodes are achieved by differential doping of thin surface layers. Overexposure and heating of these layers results in diffusion effects, which increase the work function and decrease the light sensitivity of photocathodes permanently. CCDs and CMOS detectors are much more likely to survive a moderate overexposure.

A disadvantage of solid-state array detectors is their size limitation. Because they are based on single monocrystals, they cannot exceed the sizes of the largest available wafers. Photographic plates, based on many microcrystals, can have much larger sizes. If an instrument requires a detector surface that exceeds the maximal wafer size, mosaics or "arrays of array detectors" must be constructed. Although this is possible, it requires a careful and stable alignment of the individual mosaic components, and it usually results in gaps between the individual subunits.

The size restriction of monolithic array detectors had important consequences for the design of astronomical spectrographs. The small chip sizes of early CCDs made it difficult to adapt the high-resolution coudé spectrographs of the twentieth century (designed for large photographic plates) to array detectors. Another problem was the curved focal plane of the Schmidt cameras of the classical coudé instruments. In the past, these cameras used thin, large-format photographic plates, which could be bent to approximate the curved focal surface. Obviously, this is not an option for CCDs. On the other hand, the linearity and higher dynamic range of modern array detectors simplified the use of echelle spectrographs, which (because of the variable illumination along the spectral orders) were not well suited for photography. As a result, with the development of solid-state array detectors, for high-resolution spectroscopy most large coudé spectrographs were replaced by efficient echelle instruments.

Apart from their higher sensitivity, modern array detectors have the important advantage of an essentially linear response to the incident light flux. Therefore, they are sensitive to very low light levels that cannot be recorded with the photographic process. Even more important for the spectroscopy of faint objects, however, is their linearity over a significant flux range, which allows an accurate subtraction of the sky background and of other spurious signals from the spectra. As a result of this property, it became possible to obtain images and spectra of objects whose surface brightness is much lower than the surface brightness of the night sky. Accurate sky subtraction is particularly important for observation of high-redshift objects, where, as explained in the textbooks on extragalactic astronomy (see, e.g., Appenzeller, 2009), the surface brightness decreases with the redshift z about proportional to $(z + 1)^{-4}$. Therefore, linear detectors were a fundamental prerequisite for the discovery and the present-day routine spectroscopy of high-redshift galaxies. Without the development of linear spectroscopic detectors, there would have been no chance to acquire our present knowledge of the high-redshift universe.

1.4 Extending the Wavelength Range

Prior to the invention of photography, astronomical observations were restricted to the wavelength range to which the human eye is sensitive. With the introduction of photography and photocathodes, this range was extended slightly to include the near ultraviolet above 350 nm, and the infrared at wavelengths up to about 1,200 nm. The first big step to extend the astronomically useful spectral range occurred in 1931 when Karl Jansky (1905–1950) discovered that, in addition to optical light, radio radiation reaches us from space.

1.4.1 The Radio Spectral Window

In radio astronomy, the radiation is detected and measured coherently using receivers, which record electromagnetic waves directly. Many receivers use resonance techniques and are sensitive to a distinct wavelength. The sensitivity of broadband radio receivers can be restricted to narrow frequency bands using electronic band filters. Thus, spectra can be obtained by using either several receivers designed for different frequencies or by tuning the frequency of a receiver. As a result, from the very beginning radio observations resulted in frequency information on the continua of the astronomical radio sources. Surprisingly, line spectroscopy at radio wavelengths began only about two decades after Jansky's fundamental discovery of the radio radiation from space, when Dutch astronomers measured for the first time profiles of the 21-cm hyperfine structure line of atomic hydrogen (van de Hulst et al., 1954).

The 21-cm line had been predicted by H. C. van de Hulst in 1944, and its existence had been proven by Ewen and Purcell (1951), who found an excess radiation at the corresponding frequency from the direction of the Milky Way. However, profiles of this line were not measured until 1953, when van de Hulst and his colleagues started making frequency scans around the predicted frequency with a tunable receiver. They immediately recognized the great potential of such measurements for deriving information on the rotation of the Milky Way galaxy (see Figure 1.10) and of other disk galaxies. Compared with

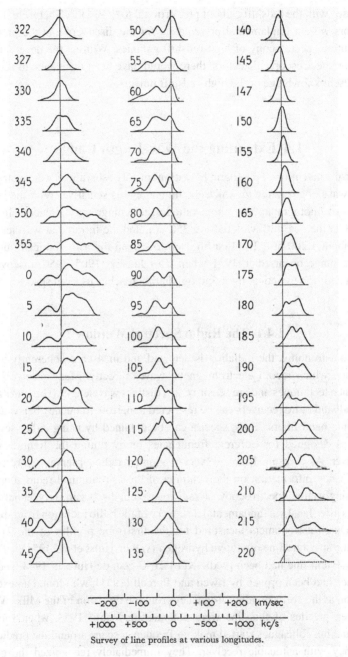

Survey of line profiles at various longitudes.

Figure 1.10. Early observations of the 21-cm line of atomic hydrogen in the Milky Way. The numbers indicate the galactic longitude in the old galactic coordinate system. From van de Hulst et al. (1954).

optical spectroscopy, such radio observations have the important advantage of not being affected by dust absorption in the galactic plane. Moreover, it was soon realized that a larger portion of the galaxy could be sampled with radio spectroscopy.

The next important milestone for astronomical spectroscopy at radio wavelengths was the discovery of hydroxyl (OH) lines in the direction of the radio source Cas A in 1963. Apart from the fact that these lines were the first interstellar molecular lines discovered at radio frequencies, it was soon recognized that one of the observed lines was the result of natural maser emission. During the following years, other intense natural masers of astronomical objects were discovered. Their intensity and narrow line widths made these radio sources a particularly valuable new tool of astronomy.

For several years, only two-atom molecules were observed in interstellar space. But starting in 1968, radio observations of dense interstellar cloud cores resulted in the discovery of complex multiatom molecules, which led to the development of the new field of *astrochemistry*. More recently, observations of carbon monoxide (CO) lines in high-redshift galaxies and quasi-stellar objects (QSOs) established an important new tool for studies of the early chemical evolution of the universe.

1.4.2 The High-Energy Sky

With the development of rockets and artifical satellites, the wavelengths of the electromagnetic spectrum, which are absorbed in the Earth's atmosphere, became accessible to astronomy and to astronomical spectroscopy. Space astronomy started in the late 1940s, when scientists of the U.S. Naval Research Laboratory (NRL) began to search for hard radiation from the Sun by launching rockets into the high atmosphere. A potential first recording of X-rays from outside the Earth's atmosphere was reported by T. R. Burnight (1949). Several times in 1948, Burnight flew packages of X-ray sensitive films on Aerobee rockets into the high stratosphere. In some cases, these films showed blackening, which, in one case, appeared to be correlated with the occurrence of a solar flare. Therefore, Burnight suggested that the films were exposed by X-rays of solar origin. Even though Burnight's results have been called in question (see, e.g., www.aip.org/history/ohilist/4613.html), there is general agreement that solar X-rays and FUV radiation were definitely detected one year later by Herbert Friedman and his group (see Friedman et al., 1951). Friedman et al. used a captured former German V2 rocket, which could fly higher, and they succeeded in determining the solar FUV and X-ray spectrum using sensitive photon counters. The great importance of the X-ray bands for astronomy became evident only in 1962, however, when Riccardo Giacconi and his team (accidentally) discovered the first strong stellar X-ray source, Sco X-1 (Giacconi et al., 1962). The first X-ray spectra of this source were again observed by Friedman's group at the U.S. NRL using proportional counters (Friedman et al., 1966). The resulting rapid development of X-ray astronomy was described authentically by Riccardo Giacconi in his Nobel lecture (Giacconi, 2003).

With very few exceptions, X-ray astronomers have always used linear photon counting detectors. Some of the first rocket flights carried Geiger counters, which provide information on the photon flux but not on the photon energies. However, soon thereafter, proportional counters and gas scintillation counters became the standard detectors. Because these detectors give a signal for each counted photon that is proportional to the photon energy (which, in turn, is proportional to the radiation frequency), the flux measurements resulted automatically in low-resolution ($\lambda/\Delta\lambda \approx 5$–$10$) spectra. Therefore, already during the first years of X-ray astronomy, valuable information on the X-ray continua could be collected. On the other hand, with the low spectral resolution of proportional counters and gas scintillation counters, spectral lines could be detected in exceptional cases only. Attempts to use high-resolution Bragg crystal spectrometers for astronomical X-ray line spectroscopy were not successful, as the efficiency of these instruments was too low to give reliable results. Lines in astronomical X-ray spectra were finally discovered in 1976, when a group at the Goddard Space Flight Center, using the OSO-8 satellite, unambiguously identified an iron emission line in the X-ray spectrum of the radio source Cas A (Pravdo et al., 1976). Equally important was the discovery a few months later of a cyclotron line in the X-ray spectrum of Her X-1 by a German group, led by Joachim Trümper, using a gas scintillation counter on a stratospheric balloon (Trümper et al., 1978).

X-ray line spectroscopy really came of age when photon-energy sensitive CCD detectors became available by about 1990. These CCDs also use the inner photoeffect (Figure 1.6, Case (a)). However, because of the keV energies of X-ray photons, energetic photoelectrons are produced, which lose their energy by producing additional conduction electrons. In this way, each keV photon can generate about 10^3 conduction electrons. Because their exact number is, on average, about proportional to the photon energy, the observed charge provides direct information on the energy (and the frequency) of the absorbed photon.

Pioneering work with X-ray CCDs was carried out on the Japanese X-ray satellite ASCA, which was launched in 1993. Its payload included four solid-state imaging (CCD) spectrometers with a spectral resolution of about 50 (Tanaka et al., 1994). The modern large X-ray observatories (such as Chandra and XMM-Newton) use more advanced CCDs (with a similar energy resolution), but also efficient grating spectrographs of different types. These up-to-date devices will be discussed in Chapter 7.

Parallel to X-ray astronomy, methods were developed to observe the sky in the even more energetic gamma rays. Like the proportional counters of the early X-ray astronomy missions, gamma-ray detectors (such as crystal scintillators) provided spectral information from the very beginning. However, for a given energy flux (in W m^{-2}), the number of recorded photons decreases inversely proportional to the photon energy. Therefore, at gamma energies, the spectral information is often limited by the statistical errors of the small numbers of detected photons.

The first evidence for the existence of cosmic gamma rays from space was found with the Explorer 11 satellite in 1961 (Kraushaar et al., 1965), although individual sources could not be determined. Later observations identified solar flares as emitters of gamma continuum and gamma line radiation, and observations with the OSO-7 satellite established the Milky Way as a major source of gamma photons. Between 1975 and 1982, a first map of the gamma-ray sky was produced with the ESA satellite COS B.[3] Another important milestone for the development of gamma-ray astronomy was the discovery of cosmic gamma-ray burst sources by military (nuclear arms control) satellites around 1970. Later observations showed that some of these objects belong to the brightest and most distant astronomical objects in the universe.

At very high (TeV) energies, the very low photon rates from cosmic sources require large detecting areas, which cannot be realized with space observatories. Therefore, starting with pioneering work at the Whipple Observatory on Mt. Hopkins, Arizona, in the 1980s, these hard gamma photons have been detected using ground-based air-shower Cherenkov telescopes. These instruments make use of the fact that TeV photons produce showers of energetic particles in the high Earth atmosphere. Because these particles are traveling with speeds exceeding the velocity of light in air, they produce flashes of Cherenkov radiation, which are recorded by the Cherenkov telescopes. (For details, see Section 7.5.)

---

[3] Detailed information on COS B and other more recent missions and observatories mentioned in this chapter can be found at the corresponding Internet sites. Links are provided by all major search engines.

1.4.3 The Last Gaps

After radioastronomy, X-ray astronomy, and gamma-ray astronomy had been established, there still existed wavelength gaps between the ground-based IR and the radio range, and between the ground-based UV and the X-rays, where astronomical observations were not yet feasible.

The IR–radio gap was closed in the 1980s by installing millimeter-wave and submillimeter radio telescopes at high-altitude sites (see, e.g., Figure 8.4), and by orbiting IR observatories, which were equipped with cold telescopes. The technologies used at these facilities were extensions of the general techniques that had been developed for the radio and optical domains. The most important IR mission was the launch of the Infrared Astronomical Satellite (IRAS) in 1983. With IRAS, it became possible for the first time to survey the mid-IR and FIR sky and to get an inventory of the bright astronomical sources that radiate at these wavelengths. Although IRAS was mainly a photometric observatory, it included a low-resolution mid-IR spectrometer, which took more than 5,000 spectra during the lifetime of the instrument.

Closing the gap between the soft X-rays and the ground-based UV was technically more complex. Early space-based spectroscopy in the near ultraviolet and far ultraviolet (NUV/FUV) (100–320 nm) range was carried out with cameras and spectrometers, which had been developed for ground-based work, and which had been adapted to satellite observations with only modest modifications. Among these early successful experiments were the NASA satellites OAO-2 (1968) and OAO-3 (Copernicus, 1972). Most important for the development of FUV spectroscopy, however, was the International Ultraviolet Explorer (IUE) satellite, which was launched in 1978. With two echelle spectrographs (together covering the wavelength range 115–320 nm) and eighteen years of successful operation, this satellite still remains the main source of our present knowledge of the FUV spectra of astronomical objects. IUE was switched off when the Hubble Space Telescope became the main instrument for this wavelength range.

More difficult have been explorations of the EUV ($\lambda < 100$ nm) wavelengths and the very soft X-rays. As pointed out in Chapter 7, the strong absorption of extreme ultraviolet (EUV) photons in most materials requires special optics and detectors for this range. The conventional spectrometer on OAO-3, although covering wavelengths down to 90 nm, had little sensitivity below 100 nm, therefore. More efficient were the EUV spectrometers of the Hopkins Ultraviolet Telescope (HUT) and of the ORFEUS FUV/EUV spectroscopy experiment, which were flown for several short-term missions between 1990 and 1996. HUT and ORFEUS resulted in a first representative sample of astronomical

EUV spectra between 90 and 100 nm, and in some data on wavelengths down to 40 nm. Many more spectra in the 90–120 nm range have been collected more recently by the Far Ultraviolet Spectroscopic Explorer (FUSE) satellite, which was operated by NASA as part of the Origins program between 1999 and 2007. Even today, however, part of the EUV spectral range remains poorly explored.

2

Spectroscopy in Present-Day Astronomy

As pointed out in the preface, this book is devoted to the observational and technical aspects of astronomical spectroscopy. Thus, a detailed discussion of the physical analysis and the use of astronomical spectra is outside the scope of this work. On the other hand, some details described in the following chapters can be understood only from the special requirements of astronomical applications. Moreover, many astronomical spectrometers have been designed for specific tasks (although they often turned out to be most useful for other applications, and for work on objects that had not yet been discovered when the instruments were planned). Even spectrometers that are designed to cover a large range of scientific problems usually work best for certain tasks and are less efficient, or totally unsuited, for specific other applications. Therefore, in the following sections, some important applications of astronomical spectroscopy are summarized briefly to provide guidelines for the technical aspects that are described in the later chapters.

2.1 Spectral Classification

In many natural sciences, the first step toward an understanding of objects or phenomena has been sorting them into classes according to their observed properties. In astronomy, the first classification occurred when early observers of the night sky started to distinguish between planets and stars on the basis of their different motions. As noted in Chapter 1, during his first spectral observations of stars, Joseph Fraunhofer found significant differences between individual stellar spectra. A first crude classification of the stellar spectra (with four classes) was devised in 1860 by G. B. Donati (1826–1873). Later in the nineteenth century, several astronomers developed increasingly complex spectral classification schemes. As described, for example, by Hearnshaw (1986),

26

initially these classifications simply reflected spectral similarities. However, it was soon realized that the spectral properties were related to differences in fundamental stellar parameters – namely the surface temperature, the total radiation power (in astronomy, denoted as *absolute luminosity*), and the atmospheric chemical composition. By correcting and improving earlier classification systems, in 1943 William W. Morgan (1906–1994) and his colleagues developed the so-called Morgan-Keenan-Kellman (MK or MKK) classification system (Morgan et al., 1943), which is still widely used today. Later, the MK system was extended and supplemented by various authors (e.g., Morgan et al., 1978; Walborn et al., 2002; Kirkpatrick, 2005; Sion et al., 1983). A comprehensive description of the MK system and its history is given in a book devoted to this topic by Gray and Corbally (2009), in which the interested reader can find many details that cannot be discussed here. However, in view of the importance of the MK classification for the statistics of stellar populations, and because MK spectral types are widely used in the astronomical literature, a summary of the main features of the MK system is given here.

MK spectral types are specified by a combination of letters and numbers. In most cases, an MK designation consists of a capital letter followed by an Arabic numeral between 0 and 9 and a Roman numeral between I and VII. A well-known example is the MK type of the Sun, which is G2V. For stars with normal (solar-like) atmospheric chemical composition, the MK designation always starts with one of the letters O, B, A, F, G, K, M, L, or T. These letters form a sequence of decreasing effective surface temperature, where type O corresponds to $T_{eff} > 33,000$ K, and type T to $\approx 1,000$ K. The Arabic numeral following the leading letter is used to subdivide the temperature classes, with the temperature decreasing from 0 to 9. The Roman numeral is used to indicate the luminosity. The luminosity class I (sometimes subdivided into Ia and Ib) defines the most luminous (or "supergiant") stars, V stands for (core hydrogen-burning) main-sequence stars, and class VII corresponds to the white dwarf stars. For stars with a peculiar chemical composition of their surface layers (such as the Wolf-Rayet stars, which have lost their hydrogen-rich outer layers), special designations are used (see, e.g., Crowther, 2007). Details on the classification of chemically peculiar stars and the meaning of various additions to the basic types (such as the lower-case letters p, e, f, n, and s) can again be found in the book by Gray and Corbally (2009).

Effective temperatures and luminosities corresponding to the MK types are listed in the compendium *Astrophysical Quantities* (Cox, 2000) and in similar collections of astronomical data. By comparing the observed flux of a star with its intrinsic luminosity derived from its spectral type, the MK system is frequently used to derive stellar distances. Distances that have been derived in this way are called *spectroscopic parallaxes.*

MK spectral types are determined by comparing observed spectra with the spectra of MK standard stars, which define the system. The standard stars are distributed over a large part of the sky to make them accessible from all major observatory sites. To get accurate results, the target stars and the standard stars should be observed with the same spectrometer under similar conditions. If this is not possible, the effects of different instrumentation must be carefully modeled and taken into account in the results.

Because the system is defined by standard stars, MK classifications can, in principle, be carried out with any spectrograph, telescope, and detector. Of course, the spectral resolution and signal-to-noise ratio must be sufficient to reliably determine the spectral differences between the different classes and subclasses. Because the original MK classes were defined using blue (or red) photographic spectra (with a resolution $\lambda / \Delta \lambda \geq 2,000$), spectral types obtained at different wavelengths should be verified with sample spectra that cover the spectral range used by Morgan and his coworkers.

Historically, the comparison between target and standard spectra has been carried out by visually inspecting photographic spectral plates through a microscope. Today, digital spectra and automatic procedures using suitable algorithms and software packages are used for spectral classification (e.g., Bazarghan and Gupta, 2008). Apart from being the only feasible option for large samples, the automatic procedures have the advantage of providing more uniform results. However, care must be taken that the corresponding software recognizes binary spectra, circumstellar emission lines, and interstellar spectral features, which could cause classification errors.

To illustrate the potential and limitations of the MK system, examples of blue (380–500 nm) photographic spectra of MK standard stars are reproduced in Figure 2.1. These spectra were obtained with the spectrograph that was used originally to define the system. The wavelength increases from left to right. Least complex are the spectra of the moderately hot A-type stars, which have surface temperatures of about 10,000 K and which, in this wavelength range, are dominated by strong hydrogen lines of the Balmer series (from Hβ to H9). In the spectrum of the A0V star, very few other lines are visible. The spectra of more luminous stars of this temperature class are similar, except for the line widths, which decrease markedly with the luminosity. In the spectrum of the B3V star (with a surface temperature $\approx$ 20,000 K) the H lines are weaker because at B3 the hydrogen is more highly ionized. In addition, the B3 spectrum shows various lines of atomic helium (HeI), which (because of their high excitation energy) are not present in the spectra of cooler stars. The strength of the hydrogen lines

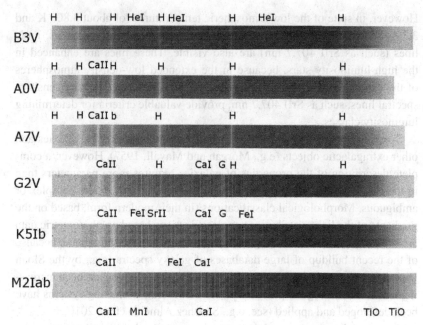

Figure 2.1. Examples of photographic spectra of MK standard stars. The spectra were obtained with the spectrograph that was used to define MK stellar spectral classification. Some of the spectral features used to determine MK types are indicated above or below the spectra. Details are described in the text.

also decreases between A0 and A7. This weakening of the Balmer lines with lower temperatures is due to a decrease of the population of the $n = 2$ energy level of atomic hydrogen. With a further temperature decrease, the hydrogen lines become inconspicuous. On the other hand, at the spectral type A7, a line of singly ionized calcium (CaII) at 393.4 nm (which is absent at B3 and weak at A0) becomes clearly visible. CaII also contributes to the feature labeled b, which at A7 is a blend of the hydrogen line Hϵ and CaII 396.8 nm. At G2V (which is the spectral type of the Sun), the two CaII lines have become the strongest spectral features in the wavelength range reproduced in Figure 2.1. In addition to CaII, many lines of atomic iron (FeI) and other iron-group elements are seen at this spectral type. Moreover, lines of atomic CaI become visible. The distinct broad feature labeled G in Figure 2.1 (called the *G band*) is a blend of iron lines and molecular features. (Its designation goes back to Fraunhofer.)

In the two cool (K5 and M2) supergiant spectra of Figure 2.1, the atomic lines of metals are even more dominant, and in the M star molecular bands of titanium oxide (TiO) (with characteristic band-head absorption steps) become important.

However, in spite of the low atmospheric temperatures (of about 3,800 K and 3,300 K, respectively) in the spectra of these luminosity class I stars, ionic lines (such as SrII 407.7 nm) are also visible. These lines are enhanced in the high-luminosity stars, because in the extended low-density atmospheres of the supergiants, recombination rates are low. Obviously, density-sensitive spectral lines, such as SrII 407.7 nm, provide valuable criteria for determining luminosity classes.

Spectral classification schemes have also been developed for galaxies and other extragalactic objects (e.g., Morgan and Mayall, 1957). However, a complete description of the properties of galaxies requires more parameters than in the case of stars. This makes their spectral classification more complex or ambiguous. Morphological classifications (in their modern form, based on the gradients of the light distribution) or classifications based on photometric data are normally preferred in extragalactic astronomy. On the other hand, because of the recent buildup of large databases of galaxy spectra (e.g., by the Sloan Digital Sky Surveys), interest in galaxy spectral classification has been increasing again, and during the past few years various new promising systems have been developed and applied (see, e.g., Sanchez Almeida et al., 2010).

An example of a spectral classification outside the optical range is the continuum classes of extragalactic radio sources. Normally, the flux F_ν of these sources decreases with the frequency ν according to $F_\nu \sim \nu^{-\alpha}$, where typically $\alpha \approx 0.8$. However, in compact sources, in which synchrotron self-absorption is important, the exponent α (called the *spectral index*) tends to be close to zero or even negative. Such objects are classified as *flat-spectrum radio sources*.

Power-law spectra are also observed at X-ray frequencies, and the spectral index can again be used to distinguish between different object classes. Examples of spectral classes in this wavelength band are the different types of X-ray-emitting pulsars and neutron stars (see, e.g., Becker, 2009).

2.2 Radial Velocities

Among the traditional tasks of astronomy is the measurement of the motions of celestial objects. A full characterization of the space motion of an object requires the derivation of velocities along three space coordinate directions. Convenient coordinates are orthogonal systems on the celestial sphere (such as right ascension [RA] and declination [Dec]) and (as a third coordinate) the direction along the line of sight (LOS) to the object in question. If the distance is known, the velocity components perpendicular to the LOS can be determined by measuring the changes of the celestial coordinates with time. However, for a given space velocity, the coordinate changes (called *proper motions*) are inversely proportional to the distance, and are easily measurable only for sufficiently close objects. The velocity component of a target along the LOS is called the *radial velocity*. This component can be determined from spectroscopic observations by means of the Doppler effect. By convention, in astronomy a radial velocity is defined as positive if the distance between the observer and the observed object is increasing. For a relative LOS velocity v between the object and the observer, special relativity predicts a wavelength change,

$$\lambda_{obs} = \lambda_0 (\frac{c+v}{c-v})^{1/2}, \tag{2.1}$$

where λ_{obs} is the observed wavelength, λ_0 the intrinsic wavelength, and c is the velocity of light. In astronomy, spectral shifts are often expressed by the dimensionless quantity $z = (\lambda_{obs} - \lambda_0)/\lambda_0$. Using this definition, we get from Equation 2.1 the *Doppler shift* z_D,

$$z_D = \frac{\lambda_{obs} - \lambda_0}{\lambda_0} = (\frac{c+v}{c-v})^{1/2} - 1. \tag{2.2}$$

By expanding Equation 2.2 in terms of v/c, one gets for $v \ll c$ the approximate relation

$$z_D = \frac{\lambda_{obs} - \lambda_0}{\lambda_0} \approx \frac{v}{c}, \tag{2.3}$$

where for $|z_D| < 0.1$ the error of z_D is $< 0.6z^2$. Thus, for small velocities, Equation 2.3 is sufficiently accurate for most astronomical applications. In practice, spectral shifts z are measured by comparing the observed wavelengths of spectral features (such as lines, absorption edges, or emission maxima) with the corresponding undisplaced wavelengths. In contrast to proper motion-based velocities, the accuracy of Doppler measurements is independent of the distance.

Among the best-known applications of radial velocity data is the derivation of orbits of binary-star components and of extrasolar planets. Apart from cases in which the orbital planes are exactly perpendicular to the LOS, orbital motions result in a periodic radial velocity variation. By measuring these periodic time variations, the orbital parameters (with the exception of the inclination angle i) can be determined from the radial velocity changes. If the inclination is known from other data, the mass of the involved objects can be estimated. In fact, most of our knowledge on stellar mass values and on the mass of extrasolar planetary systems is based on such radial velocity measurements. With

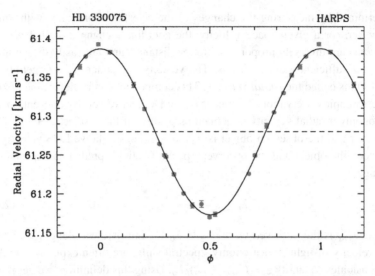

Figure 2.2. Observed radial velocity variations of a star with a planetary companion. The solid line shows the velocity predicted from the calculated orbit. The mean deviation of the observed velocities from the predicted curve is about 2 m/s only. From Pepe et al. (2004).

present-day techniques radial velocity variations as small as about 1 m/s can be observed. An example of a precise measurement of this type is given in Figure 2.2.

Another mechanism that produces periodic radial velocity variations in stellar spectra is the periodic radius change in pulsating variable stars. The evaluation of such data is an important source of information on stellar radii.

On larger scales, spectral radial velocity measurements of stars, of integrated star light, and of interstellar emission and absorption lines (observed at optical or radio wavelengths) are used to map the velocity fields of galaxies and to determine the distribution of their visible and dark mass. On even larger scales, the radial velocity dispersion of the members of galaxy clusters can be used to measure the (mainly dark) matter content of the largest gravitationally bound systems in the universe.

2.3 Gravitational and Cosmological Redshifts

Einstein's theory of gravitation (general relativity) and cosmological world models based on this theory predict that large spectral shifts can be observed in the spectra of astronomical objects even when the sources are at rest. Important examples are the redshift of radiation emitted in deep gravitational potential wells and the light from distant galaxies and QSOs, which is redshifted as a result of the expansion of our universe.

2.3.1 Gravitational Redshifts

For stars (and other spherical objects), general relativity predicts a gravitational redshift, which is given by the simple relation

$$\lambda_{obs} = \lambda_0 (1 - \frac{2GM}{rc^2})^{-1/2}, \tag{2.4}$$

where λ_{obs} is the observed wavelength, λ_0 the intrinsic wavelength, G the gravitational constant, M the objects mass, c the velocity of light, and r the radius where the light is emitted. For sufficiently large radii r, Equation 2.4 can be approximated by

$$\lambda_{obs} \approx \lambda_0 (1 + \frac{GM}{rc^2}). \tag{2.5}$$

If one inserts into Equation. 2.4 the mass and radius of the Sun, one finds a relative wavelength change of about 2.12×10^{-6}; that is, a gravitational redshift for the solar spectrum corresponding to a Doppler velocity of about 600 m/s. Radial velocity measurements in the solar spectrum indeed give the predicted redshift for some spectral lines. However, other lines show different (usually larger) shifts, and the spectral shifts tend to vary over the solar disk. The origin of the additional nonrelativistic line shifts are convective motions in the solar atmosphere. As the example of the Sun shows, gravitational redshifts are present and measurable in stellar spectra, but they are often difficult to separate from additional, nonrelativistic effects. Moreover, to determine the gravitational shift, the radial velocity (and the corresponding Doppler shift) of the star must be known exactly. Normally, this value is known with sufficient accuracy only for binary systems, for which the mean (or systemic) velocity can be used as a reference.

Large gravitational shifts are observed in the case of white dwarf stars and neutron stars. White dwarfs (WDs) have masses comparable to the solar mass, but radii that are typically only about 1/100 of the solar radius. Hence, according to Equation 2.5, their relativistic wavelength changes are larger by a factor of the order 10^2. Unfortunately, although the wavelength shift is larger than in normal (nondegenerate) stars, accurate measurements are more difficult. One problem is that (as pointed out earlier) accurate line shift measurements are possible only if the WD is a member of a binary system. However, because of their small radii, WDs in binary systems are normally much fainter than their companions. As a result, the spectra of WDs in binary systems are often heavily contaminated by light of the primary component. Moreover, because of their high atmospheric pressure, WDs have very broad lines. Both effects impair accurate wavelength measurements. The existing derivations of WD gravitational redshifts are again consistent with the prediction of Equation 2.4, but not precise enough to provide new astrophysical information.

Not surprisingly, the highest gravitational line shifts (up to $z = 0.35$; see Cottam et al., 2002) have been reported for neutron stars, which have radii of only a few kilometers. However, our poor understanding of the line emission regions of neutron stars makes it difficult to obtain reliable conclusions from these measurements. More accurate, and physically much more useful, information on the strong gravitational fields near neutron stars can be collected by means of pulsar timing studies, in which various techniques initially developed for spectroscopy have been successfully adapted (e.g., Colpi et al., 2009). Because of the highly constant intrinsic frequency of the radiation pulses from rotating neutron stars, their frequency shifts can be measured with much higher accuracy than in the case of (always naturally broadened) spectral lines.

2.3.2 Cosmological Redshifts

Astrophysically more important than gravitational redshifts of spectral lines are the *cosmological redshifts* that are caused by the expansion of the universe. As explained in the textbooks on extragalactic astronomy and cosmology (e.g., Longair, 2008), for objects that are locally at rest, the corresponding wavelength change is given by

$$\lambda_{obs} = \lambda_0 \frac{a(t_0)}{a(t_{LE})}, \tag{2.6}$$

where $a(t_0)/a(t_{LE})$ is the ratio between the present-day cosmic scale factor $a(t_0)$ and the scale factor $a(t_{LE})$ at the time of the light emission t_{LE}. The function $a(t)$ can be calculated numerically if the cosmic parameters are known. For certain cosmic epochs, the function $a(t)$ can also be approximated by analytic expressions. Analogous to Equation 2.2, the wavelength shift is often expressed by the dimensionless quantity z_c, which is defined as

$$z_c = \frac{\lambda_{obs} - \lambda_0}{\lambda_0} = \frac{a(t_0)}{a(t_{LE})} - 1. \tag{2.7}$$

However, for historic reasons, in the astronomical literature, instead of z_c sometimes the product cz_c is given, where c is the vacuum velocity of light. This product obviously has the dimension of a velocity, but does not correspond to a real velocity. Therefore, for cz_c we can have (and often do have) $cz_c > c$, and Equations 2.1 and 2.2 must not be applied to cz_c. On the other hand, as z_c depends on the "look-back time" $t_0 - t_{LE}$, it is a convenient measure of cosmic distances. When using cz_c as a distance measure, one must keep in mind that Equation 2.6 is strictly correct only for objects that are locally at rest. Therefore, in practice, observed redshifts must be corrected for the motions of the observed objects (and the motions of the solar system and of our galaxy) before they can be used to characterize distances.

2.4 Astrophysical Applications

Astrophysics normally is defined as the branch of astronomy that investigates the physical structure and the physical processes of astronomical objects. Because of its role as a remote sensing tool, spectroscopy is particularly valuable for these tasks, and the analysis of spectra is one of the main activities of astrophysicists. As a result, there exists an extensive literature on the astrophysical applications of spectroscopy. Timely general introductions to the use of spectra in astrophysics are presented, for instance, in the books of Pradhan and Nahar (2011) and of Tennyson (2011). A good description of the physical interpretation of radio spectra has been given by Wilson et al. (2009). Therefore, in the following sections only a few important examples will be discussed.

2.4.1 Stars

Studies of the atmospheres of stars (including the Sun) marked the beginning of astrophysics in the nineteenth century, and stellar astrophysics still remains one of the important applications of astronomical spectroscopy. Except for the neutron stars (which may be hotter), stellar surface temperatures vary between about 10^3 and a few times 10^5 K. Therefore, stars radiate mainly between the MIR and the EUV, and stellar spectroscopy is essentially confined to the optical spectral range.[1] Because the temperature in the visible layers of stars decreases outward, stellar line spectra are usually dominated by absorption lines (see Figure 2.1).

The main objective of stellar spectroscopy is to derive the physical properties of stellar atmospheres. For this purpose, the observed spectra are compared with synthetic spectra, which are calculated using theoretical atmospheric models.

---

[1] In this book, the wavelengths for which conventional optical techniques can be used will be referred to as the *optical range*. This includes the FUV, UV, visual, and IR bands.

Suitable computer codes have been developed by Kurucz (1970), Lester and Neilson (2008), Jack et al. (2009), and many others. Some of these codes are publicly available. Instead of modeling whole spectra, it is sometimes of advantage to simulate certain spectral features separately. Modeling the continua usually gives the most reliable temperature estimates, whereas line profiles give direct information on the rotation, stellar winds, the surface gravity, and the presence and strengths of magnetic fields.

Neutron stars typically have surface temperatures near 10^6 K. Thus, their thermal spectra must be observed at X-ray wavelengths. Because of their small diameters, neutron stars are weak thermal sources, but studies of their surface emissions have become possible because of the high sensitivity of the current X-ray observatories.

2.4.2 Interstellar Gas and Dust

The interstellar space of our galaxy is filled with a low-density gas or plasma with temperatures ranging between about 10 K and $>10^6$ K. Because of the low density of the interstellar medium, the rich emission and absorption line spectrum of the interstellar gas includes "forbidden" emission lines, which cannot be produced under terrestrial conditions. Depending on the temperature, interstellar lines occur at wavelengths anywhere between the radio range and the X-rays. Thus, comprehensive studies of the interstellar medium require spectroscopy over the whole wavelength range that is accessible to present-day astronomy.

The mean free path of photons varies greatly between different interstellar lines, ranging from optically thin to completely opaque conditions. Moreover, because these lines, which are formed in a rarefied and often cool gas, are intrinsically narrow, their actual line widths and their radiative transfer are strongly affected by the interstellar velocity fields. As a rule, high-resolution spectra are required to determine and evaluate their velocity structure.

The methods of analyzing the individual interstellar lines differ significantly. Particularly important for studies of the interstellar medium are the *forbidden lines*, whose transition probabilities are too small to be observable under laboratory conditions. Usually (but not always), saturation effects and self-absorption are absent in these lines. As a result, the strength of the forbidden emission lines is often directly related to the abundance of the corresponding atoms and ions. Accurate spectroscopy of these lines, therefore, forms the most reliable basis of chemical abundance derivations in the universe. Comprehensive general descriptions of the methods that are used to analyze interstellar line spectra can be found in the textbooks of Lequeux (2005) and Dopita and Sutherland (2003).

About 1 percent of the interstellar matter of our galaxy is present in the form of small (typically $\leq 1\mu m$) solid particles, which are called *interstellar dust grains*. Because the grains tend to evaporate at higher temperatures, they usually have temperatures $\leq 10^3$ K. As a result, the interstellar dust grains always radiate at IR and submillimeter wavelengths. Apart from an approximate black-body continuum, they emit (and absorb) continuous spectra with characteristic solid-state bands. These bands provide valuable information on the chemical composition of the grains and on the conditions in interstellar space. Although studies of the grain emission require IR spectroscopy, dust absorption bands are observed at IR, visual, and even UV wavelengths. For details, see again Dopita and Sutherland (2003).

2.4.3 Accretion-Powered Sources

Stars are defined as objects that are in hydrostatic equilibrium. To retain their thermal and hydrostatic equilibrium in spite of the energy loss by radiation, they must produce heat in their interiors. In most stars, the corresponding energy sources are nuclear reactions. Exceptions are the white dwarfs and the neutron stars (which radiate by consuming their internal thermal energy) and pre–main-sequence stars (which produce heat by slowly contracting and compressing their gas).

In addition to the hydrostatic stars, galaxies include objects that are powered by the infall of matter onto a stellar core or into a black hole. This matter either originates from a companion star or from the surrounding interstellar space. Because the infalling matter always contains angular momentum, normally it cannot fall directly onto a stellar surface or into a black hole. Instead, the infalling matter accumulates in an *accretion disk*, in which the gravitation of the central object and orbital centrifugal forces compensate each other. Internal friction in the disk results in a transfer of angular momentum, an inward movement of the matter, and a heating of the disk. An accretion disk can extend all the way to the surface of the central object, or it can be truncated by instabilities or magnetic fields. In this case we observe (quasi) free-falling matter inside the truncation radius.

Examples of accretion-powered radiation sources are protostars, young stellar objects that are still accumulating matter, and close-binary components that accrete mass that has been ejected by a companion.

Because of their dusty envelopes, protostars can be observed only at IR and radio wavelengths. Mass accreting pre–main-sequence objects (the "classical T

Tauri-stars") can be observed over a broad spectral range, extending from radio wavelengths to X-rays. Because T Tauri accretion disks are truncated by magnetic fields many stellar radii above the central cores, the inflow to these objects reaches free-fall velocities. The resulting strong shock fronts (at which most of the accretion energy is generated) radiate X-ray and UV continua, which in part are absorbed and reprocessed by the gas of the stellar surface, by the accretion flow itself, and by the gas and dust of the outer disk. The continua of the accretion shocks often strongly dominate the spectra. As a result, the absorption spectra of the cool underlying photospheres tend to be weak or invisible (see, e.g., Figure 2.3) Additional spectral components of the complex T Tauri spectra are visual and UV emission lines excited by the accretion shocks, and lines originating in winds and mass outflows from the disk and the central stars. Some of these outflows, occurring at supersonic speeds, are well collimated toward the rotation axes of these systems, and are called *jets*. Jets and winds are important components of mass-accreting systems because they carry away the angular momentum that otherwise would halt the inflow to central objects.

In the case of mass-accreting binary components, the observed spectra depend on the physical nature of the accreting central object. From energy conservation follows that for a fixed accretion rate and a given central mass, the energy production is inversely proportional to the radius at which the accretion power is generated. Thus, accretion disks and shocks near compact objects such as white dwarfs, neutron stars, and black holes are particularly luminous sources. Because of their high luminosity and small size, accretion disks around compact sources also must be particularly hot, with spectra typically peaking in the X-ray or EUV range. As a result, X-ray spectroscopy plays a most important role in studying these objects. However, as in the case of the T Tauri stars, part of the X-ray flux is often absorbed and converted into UV and visual line radiation. Therefore, observations in the optical wavelength bands are critical, too. Because of their broad spectra, studies of mass-accreting stars typically require spectroscopic data extending from the X-ray to the radio range. However, as illustrated by Figure 2.3, high-resolution optical (UV to NIR) spectra play a central role, as they provide detailed information on the accretion flows and winds.

The most spectacular accretion-powered objects are the "supermassive" (about $10^5 < M/M_\odot < 10^{10}$) black holes in the cores of massive galaxies. During periods of high accretion rates, their luminosity can exceed the total brightness of the integrated star light of their host galaxies by large factors. In this case, we observe a *quasi-stellar object* (QSO). At somewhat lower (but still high) accretion rates we observe an active galactic nucleus (AGN) in the core of a galaxy. Similarly to the accretion-powered radiation sources with

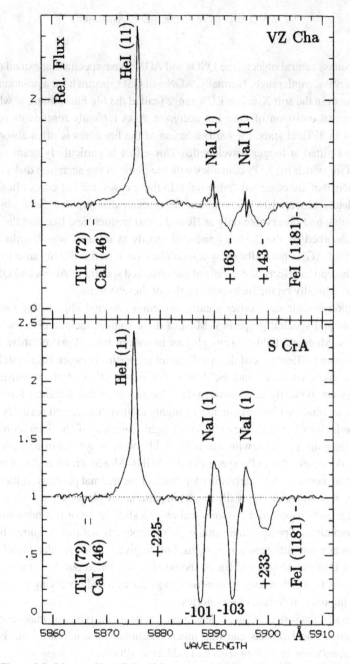

Figure 2.3. Line profiles of the NaI resonance doublet of two mass-accreting young stellar objects (VZ Cha and S CrA). Both objects show redshifted accretion-flow absorption components with the velocities (in km/s) indicated. S CrA also shows prominent wind absorption features (with $v \approx -102$ km/s). Other accretion signatures in this spectrum are the exceedingly weak photospheric lines (of TiI, CaI, and FeI) and the strong HeI emission originating from the accretion shock. From Krautter et al. (1990).

stellar-mass central objects, the QSOs and AGNs have spectra that extend over a large wavelength range. Normally, AGN and QSO spectra have a pronounced maximum in the soft X-ray or EUV range (called the *big blue bump*), at which the thermal emission of their hot accretion disks typically reaches its peak. But, as in T Tauri stars, the hard radiation of the hot cores is often absorbed and re-emitted at longer wavelengths. This effect is particularly dramatic in AGNs for which the LOS coincides with the plane of the accretion disks (i.e., in AGNs that are observed "edge-on"). In these cases, the hot cores often are completely obscured by the cool and dusty outer disk regions. Such objects are visible and observable only at IR and radio frequencies. Because the UV light absorbed by the dust is reradiated mostly at infrared wavelengths, the spectra of AGNs normally show a secondary peak in the IR. Because of the disk absorption effects, the details of the observed spectra of AGNs and QSOs depend critically on the inclination angles of these systems.

Although their sizes differ greatly, accretion-powered objects share many physical and spectral properties. Most of these sources include disks, winds, and jets. Magnetic fields always play an important role. However, there also exist marked differences of the wind and jet properties between (most) stellar-mass accretion sources and the AGNs. Whereas outflows from stellar-mass sources are typically supersonic with velocities of a few hundred km/s, in AGNs we observe plasmas containing highly relativistic charged particles and flow velocities close to the velocity of light. Because of the interaction of the relativistic particles with magnetic fields, we get synchrotron emission, which dominates the radio spectra of the AGNs. Moreover, inverse Compton scattering between the relativistic particles and the thermal photons emitted by the hot accretion disks typically results in gamma-ray photons.

All accretion-powered astronomical objects show more or less pronounced time variations. Spectroscopic studies of such objects not only require observations at many different wavelengths, but, to give reliable results, the observations in different spectral bands should also be carried out simultaneously. Often this is possible only by means of large coordinated observing programs involving many different observatories.

Readers interested in more details on the different types of mass accreting astronomical objects can find much additional information in the book *Accretion Power in Astrophysics* by Frank et al. (2003).

2.4.4 Galaxies

The luminous matter of galaxies includes all the object types mentioned in the preceeding paragraphs – stars, black holes, interstellar gas, and interstellar dust. As a result, galaxy spectra are a combination of the spectra described in the previous paragraphs. However, the relative contributions of the different components differ greatly between different extragalactic objects. An extreme example already mentioned is the QSOs, whose spectra are completely dominated by an accretion-powered central source. Supermassive black holes are present in at least a majority of the more massive galaxies. However, in most cases, the accretion rates are too small to provide a significant contribution to the total radiation flux. Therefore, in most cases, galaxy spectra are dominated by the light of the stars. The details of the stellar contribution to the galaxy spectra depend on the age of the stellar populations. Galaxies with only old (>1 Gyr) stellar populations (such as most elliptical galaxies) contain mainly cool stars with red and NIR continua. Galaxies with young populations (such as spiral galaxies and blue irregular galaxies) contain (in addition to red stars) hot blue stars, which, owing to their high luminosity, dominate the UV and visual spectra. Particularly young stellar populations are found in high-redshift galaxies, which are observed at early cosmic epochs, where the stars simply did not yet have the time to become old.

The spectral contributions of the interstellar matter to galaxy spectra is again closely related to the age of the stellar populations. Galaxies with young stars always contain cool interstellar matter, which is the raw material from which new stars are being formed. As in our Milky Way galaxy, the cool interstellar gas has a rich molecular, atomic, and ionic absorption and emission spectrum. In addition to spectral lines originating in the cool interstellar gas, in star-forming galaxies we also find emission lines that are produced in moderately hot ($\sim 10^4$ K) gas clouds, which are ionized by the UV radiation of the hot young stars. In elliptical galaxies, which contain old stars only, we find no or only little cool or warm interstellar matter. Instead, the interstellar gas in massive elliptical galaxies usually is a hot ($> 10^6$ K) and highly ionized plasma, which radiates mainly at X-ray wavelengths.

2.4.5 Intergalactic Space and Background Radiations

Although the density is much lower than in the interstellar medium of galaxies, the space between the galaxies is not empty, but filled with highly ionized warm or hot gas. In spite of its low density, this gas contains most of the baryons of the universe. A significant fraction of the intergalactic gas is concentrated in massive galaxy clusters, in which it is in hydrostatic equilibrium with the clusters' gravitational fields. The temperature of this intercluster gas is typically $> 10^7$ K. Therefore, this gas radiates mainly keV X-rays. The observed X-ray spectra provide important information on the temperature distribution and the mass of this gas, and indirectly also on the dark matter that dominates the gravitational potential of the galaxy clusters.

Outside galaxy clusters, the gas is less hot and (even) less dense. Similar to the intercluster gas, the general intergalactic gas is highly ionized. However, a small fraction consists of atomic hydrogen and of atoms and ions of other common elements. These atoms produce measurable Lyman-line and Lyman-continuum absorption of hydrogen, as well as resonance absorption lines of various other elements. Because (owing to the cosmic expansion) at earlier cosmic epochs the matter density was higher and the ionization of the intergalactic gas was lower, the intergalactic line absorption is particularly strong in the spectra of high-redshift objects.

The atomic component of the intergalactic matter is known to have a filamentary structure. Therefore, the line absorption occurs at discrete redshifts and a single spectral line (such as Lyα) can produce numerous discrete absorption features (corresponding to different redshifts) in the spectra of very distant objects (see, e.g., Appenzeller, 2009). Because of their strength, the Lyα lines normally dominate the intergalactic spectra with many sharp lines. Therefore, such spectra are referred to as the *Lyman-alpha forest*. Although the rest wavelength of Lyα is 121.57 nm, for redshifts $z \geq 2$ this "line forest" is redshifted into the visual spectral range, where it can be studied conveniently with ground-based optical spectra. These ground-based observations form the basis of most of our present knowledge of the intergalactic medium.

The hot intergalactic gas is expected to produce a (more or less smooth) background at X-ray wavelengths. However, deep X-ray observations in small test fields show that most of the observed X-ray background is caused by unresolved discrete sources, such as AGNs (see, e.g., Brandt and Hasinger, 2005). Unresolved discrete sources also appear to provide the main contribution to the backgrounds observed at most other wavelengths. An important exception is the cosmic microwave background (CMB), which has its peak intensity in the high-frequency radio range near 150 GHz (corresponding to a wavelength of about 2 mm), and which is the most intense background radiation outside galaxies. The CMB was emitted when the (originally completely opaque) universe became transparent to its dominant radiation field about 380,000 years after the Big Bang. Its energy distribution is an almost perfect black-body spectrum with a radiation temperature of 2.73 K. The deviations of the observed spectrum from a perfect black body, and the variations of the radiation temperature over the sky provide important constraints on various cosmological parameters (see, e.g., Longair, 2008).

Apart from providing information on their nature and their origin, the spectra of the cosmic background radiations form an important tool for separating the different background components and for distinguishing between backgrounds and foreground sources. As an example, we note that spectroscopy is particularly efficient to separate foreground objects from the CMB because there exists no other natural source of black-body radiation of such a low temperature, and because most discrete radio sources have distinct nonthermal spectra.

2.5 Magnetic Fields

The properties of many astronomical objects depend on magnetic fields. In most cases, our knowledge about the presence and strength of these fields is based on spectroscopic observations. The field strengths that have been measured in astronomical spectra extend from 10^{-9} Tesla to $> 10^8$ Tesla (or 10^{-5} to $> 10^{12}$ Gauss). Naturally, different methods must be employed to cover this huge range of field strengths.

Measurements of magnetic fields in the interstellar medium and in most stars generally make use of the Zeeman effect. This effect (which is explained in detail in textbooks on atomic physics) describes the interactions of the magnetic fields of the electrons and electron orbits in atoms with weak external magnetic fields. If atoms and molecules with a nonvanishing total angular momentum experience an external field, the angular momentum starts to precess around the direction of the external magnetic field. The angles to the external field are quantized with a magnetic quantum number M. If the external magnetic field is absent or negligible, M has no effect on the atomic energy levels. If a magnetic field is present, each energy level corresponding to a given value of the total angular momentum, characterized by a quantum number J, splits into $2J + 1$ sublevels, according to

$$E_m = E_0 + \frac{e}{4\pi m_e c} g_L B M, \qquad (2.8)$$

where e and m_e are, respectively, the absolute charge and mass of the electron, c is the velocity of light, B is the magnetic field strength, M is the magnetic quantum number, and g_L is the so-called Landé factor. M is zero or an integer $\leq J$. Thus, for $J = 1$ the quantum number M can have the values -1, 0, or $+1$. For simple atoms, g_L can be calculated from the quantum numbers of the level involved by means of an analytic formula. For more complex atoms, it must be determined empirically by laboratory spectroscopy.

In addition to splitting the atomic energy levels, the new quantum number M results in additional selection rules for transitions between the energy levels. For permitted transitions we must have $\Delta M = 0$ or ± 1. Spectral lines that correspond to transitions $\Delta M = 0$ are linearly polarized parallel to the magnetic field. Transitions with $\Delta M = \pm 1$ result in a circular polarization in a plane perpendicular to the external field direction, with the sign of ΔM determining the sign of the helicity. The combination of the level splitting and the selection rules results in a corresponding splitting of the spectral lines of atoms and molecules in magnetic fields. Because of the different quantum numbers and Landé factors, not all lines will split, and the energy difference between the subcomponents is different for the individual split lines.

Because according to Equation 2.8 the splitting of the energy levels is proportional to the field strength, the Zeeman effect can be used to measure magnetic fields with high precision, if the Landé factors can be calculated or measured. For many energy levels, g_L values are listed in the databases for atomic data (e.g., of the U.S. National Institute of Standards and Technology, www.nist.gov).

The most simple application of the Zeeman effect is the measurement of the LOS (or longitudinal) component of an ordered magnetic field by observing circularly polarized line components. As explained earlier, energy sublevels corresponding to magnetic quantum numbers M with different signs have opposite helicity and are displaced in opposite directions. If we describe the intensity of the circularly polarized light by means of the Stokes parameter V, the components corresponding to different helicities have different signs. Therefore, we get in the circularly polarized light characteristic line profiles as shown in Figure 2.4. By comparing the observed profile with the prediction of the Zeeman effect, the effective longitudinal field component can be derived. In the example of Figure 2.4, showing the measurement of the magnetic field in a cool interstellar cloud, a longitudinal field strength as low as 10^{-7} Tesla could be reliably detected and measured.

As noted earlier, a V profile as plotted in Figure 2.4 requires an ordered magnetic field. In a random field, the polarized components of different helicity tend to overlap and (because they have different signs) to cancel each other, and no polarization is observed. However, the magnetic field still results in a broadening of the line profile observed in the total intensity (i.e., in the Stokes parameter I). The amount of this broadening depends on the sensitivity of the individual lines on the magnetic field. Thus, in this case, magnetic field strengths can be derived – for example, by comparing the widths of lines with different Landé factors.

Even if the line width differences are not directly measurable, stellar magnetic fields can sometimes be derived from comparing the equivalent widths of magnetically sensitive lines with atmospheric models calculated without magnetic fields. Because the unresolved Zeeman components in the wings of lines change the saturation behavior of absorption lines, they affect the equivalent

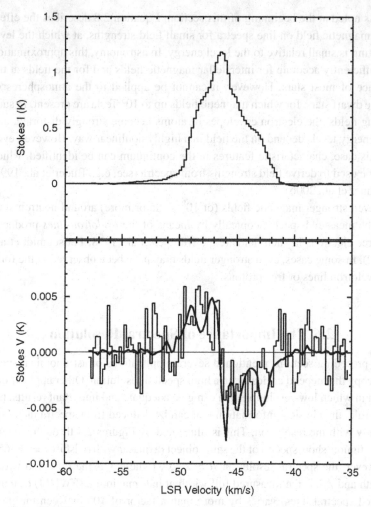

Figure 2.4. The longitudinal Zeeman effect observed in hyperfine components of the 113 GHz (0–1) line of CN in the cool interstellar cloud W3OH. Plotted are the Stokes parameters I (representing the total intensity) and V (representing the circularly polarized intensity). As customary in radio astronomy, the intensities are expressed as brightness temperatures as a function of the wavelength (expressed as a Doppler velocity). The histograms give the observed data, averaged over seven hyperfine components; the thick line shows the expected profile for an LOS magnetic field of 1.10×10^{-7} Tesla. From Falgarone et al. (2008).

widths in a way that again depends on the Landé factors. The analysis of the equivalent widths of magnetically sensitive lines has been used extensively for deriving the surface fields of T Tauri stars and other young stellar objects. For details and further literature on this method, see Johns-Krull (2007).

As noted at the beginning of this section, Equation 2.8 describes the effect of a magnetic field on line spectra for small field strengths, at which the level splitting is small relative to the level energy. In astronomy, this approximation is sufficiently accurate for interstellar magnetic fields and for the fields at the surface of most stars. However, it cannot be applied to the atmospheres of white dwarf stars, for which magnetic fields up to 10^5 Tesla are present. In such strong fields, the electron envelopes of atoms become strongly distorted, and the energy levels depend on the field in a highly nonlinear way. However, even in this case, characteristic features in the continuum can be identified, which can be used to derive field strengths from spectra (see, e.g., Ruder et al., 1994; Euchner et al., 2006).

Even stronger magnetic fields (of 10^8 Tesla or more) around neutron stars can be measured spectroscopically by means of the *cyclotron lines* produced by free electrons in such fields (e.g., Trümper et al., 1978; Pottschmidt et al., 2012). In some cases, even stronger fields may have been observed in the form of cyclotron lines of free protons.

2.6 The Importance of Spectral Resolution

The preceding sections mentioned several applications of astronomical spectroscopy that depend critically on a high spectral resolution. Other applications exist in which low-resolution data can give adequate and important results, but as a rule, the physical information that can be deduced from spectra increases greatly with the resolution. This is illustrated by Figures 2.5 through 2.7. All three figures show spectra of the same object (a quasar with redshift $z = 3.365$). However, the spectral resolution $R = \lambda/\Delta\lambda$ (where λ is the observed wavelength and $\Delta\lambda$ is the measured full width at half maximum (FWHM) of unresolved spectral lines) varies by more than a factor of 100 between the three spectra.

The spectrum in Figure 2.5, although of very low resolution ($R \approx 350$), shows sufficient details to identify the object unambiguously as a high-redshift quasar. Several broad emission lines of hydrogen, helium, carbon, oxygen, and silicon provide information on the object's chemical composition and on its redshift. The spectrum also indicates the presence of continuum absorption shortward of 4,000 Å and line absorption shortward of 5,300 Å. However, owing to the low resolution, none of the absorption lines (or their redshift) can be identified. On the positive side, the single low-resolution spectrum, based on one short exposure, shows the basic spectral properties over the whole wavelength range, which can be recorded with a silicon-based CCD detector.

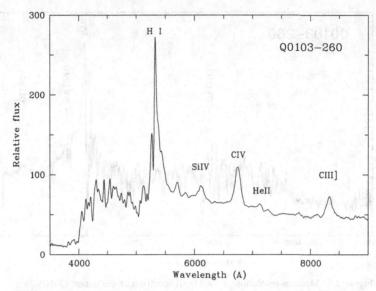

Figure 2.5. Low-resolution ($R \approx 350$) spectrum of the quasar Q0103-260. Courtesy ESO/FORS Team.

The spectrum in Figure 2.6 ($R \approx 2,000$) was obtained with the same spectrometer, but with a different grating. It shows that the absorption shortward of 5,300 Å is due to a large number of narrow absorption lines. Moreover, narrow absorption lines now are detected also at longer wavelengths. Some of the narrow absorption lines (and additional broad emission features) can be clearly identified in the medium resolution spectrum. However, short of 5,300 Å, no real continuum is apparent. This suggests that the narrow absorption features at these wavelengths actually are blends of unresolved even narrower lines.

All these lines are fully resolved in Figure 2.7. Plotted in this figure is a small (50 Å) section of a high-resolution ($R = 40,000$) spectrum obtained with an echelle spectrometer. Using the information of the full high-resolution spectrum, all narrow lines in the plotted wavelength interval can be identified unambiguously. Most of the conspicuous absorption features visible in Figure 2.7 are Lyα lines of hydrogen at different redshifts ($2.8415 < z < 2.8826$). These lines are formed by hydrogen atoms in gas clouds on the line of sight between us and the quasar. Some of these hydrogen lines reach zero rest intensity at their centers and have saturated, boxlike profiles. At these wavelengths the intergalactic medium between the quasar and the Milky Way is completely opaque.

Whereas the unsaturated hydrogen lines all have similar line widths, Figure 2.7 also contains absorption lines that are much narrower. These lines

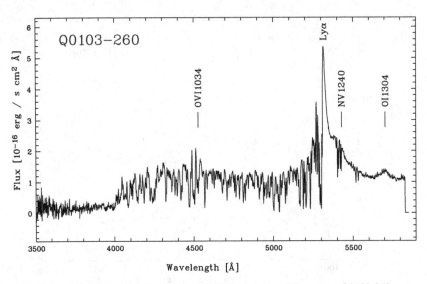

Figure 2.6. Medium-resolution ($R \approx 2,000$) spectrum of the quasar Q0103-260. Courtesy ESO/FORS Team.

(labeled FeII and SiIV) are due to singly ionized iron (FeII) at redshifts of $z = 0.9744$ and $z = 0.9737$ and SiIV at $z = 2.375$. The FeII and SiIV lines also originate in intergalactic gas on the LOS to the quasar. They are much narrower than the hydrogen lines, because in all cases the line width is determined

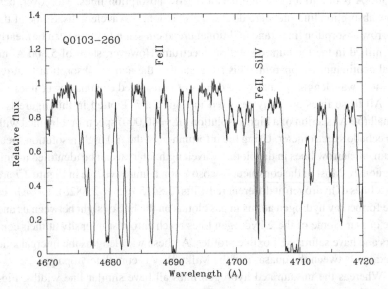

Figure 2.7. High-resolution ($R = 40,000$) spectrum of the quasar Q0103-260. Figure based on Frank et al. (2003).

by the thermal motions of the corresponding atoms and ions. Compared with that of hydrogen, the Si and Fe atomic nuclei are more massive by factors of 28 and 56, respectively. Therefore, in thermal equilibrium, the ions of Fe and and Si move much more slowly than the light hydrogen atoms. Obviously, this theory also predicts that the SiIV lines should be broader than the FeII lines. A careful measurement of the lines in the high-resolution spectrum confirms this prediction.

From the preceding paragraphs it is clear that the information content of spectra increases with the spectral resolution. For certain applications, high spectral resolution is indispensable. On the other hand, to get a similar signal-to-noise ratio, an increased spectral resolution requires more observing time. In the case of the spectra reproduced previously, the high-resolution spectrum of Figure 2.7 required about thirty times more observing time as the spectrum of Figure 2.5. Moreover, whereas the two first spectra were taken with a multiobject spectrometer, which obtained spectra of more than twenty different objects simultaneously, the high-resolution spectrum was taken with a single-object spectrograph. Taking this multiplex gain into account, the high-resolution spectrum required about 600 times more observing time. As this example shows, although high spectral resolution practically always provides additional information, carrying out observations with a higher spectral resolution than is required for a given scientific objective is inefficient and wastes valuable and expensive observing time.

3

Basic Physics of Spectral Measurements

This chapter is devoted to the physical effects and methods that are available to measure electromagnetic flux as a function of its frequency or its wavelength. The content of the present chapter will form the basis for the following chapters, in which the technical realization and the practical use of these methods will be described.

In the literature, the term *light* is often reserved for radiation at visual or optical wavelengths. However, there exists no qualitative difference between the basic properties of electromagnetic waves of different frequencies. Therefore, in the following text, the shorter term "light" will be used for all types of electromagnetic radiation.

3.1 Electromagnetic Radiation

That light is composed of electromagnetic waves has been known since the nineteenth century. Along their paths these waves produce variations of an electric and a magnetic field, which are periodic in time and space. A complete treatment of the theory of electromagnetic waves can be found in the textbooks on electrodynamics and on optics. (For a concise introduction to the subject see, e.g., Chapter 2 of Wilson et al. (2009).) In the present chapter, some of the optical effects and important relations that are essential for practical spectroscopy will be summarized briefly.

3.1.1 Monochromatic Plane Waves

We first consider the special case of a monochromatic (or single-frequency) plane light wave propagating in a nonconducting medium in the positive x direction. Because the electric and magnetic fields are coupled, a light wave

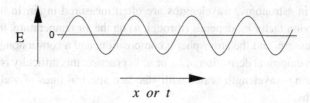

E 0

x or t

Figure 3.1. Schematic representation of a monochromatic, plane light wave.

can be characterized by either its electric or its magnetic component. In the following text we discuss the electric field only. Moreover, we look at only one of the (two) scalar components of the electric field vector.

With the preceding assumption, we can describe the periodically variable electric field of the light wave by the scalar function $E(x, t)$ according to

$$E(x, t) = E_0 \cos(kx - \omega t + \varphi), \tag{3.1}$$

where E_0 is the oscillation amplitude, $k = 2\pi/\lambda$, $\omega = 2\pi\nu$, x is a space coordinate, t is the time, λ is the wavelength, ν is the frequency of the oscillations, and φ a phase angle, which depends on the zero point of the time axis. By choosing a suitable zero point of t (or x), φ can be eliminated from Equation 3.1. A graphic representation of this type of wave is given in Figure 3.1. From Equation 3.1 it is obvious that a certain phase of the wave (e.g., the maximum of E, where, for $\varphi = 0$, we have $kx = \omega t$) travels through a medium with the *phase velocity* c_m, where

$$c_m = \frac{\Delta x}{\Delta t} = \frac{\omega}{k} = \nu\lambda. \tag{3.2}$$

The theory of electrodynamics shows that c_m can be expressed as $c_m = c/n$, where c is the velocity of light in vacuum and n (called the *index of refraction* or *refractive index*) is a function of the electric and magnetic properties of the material through which the wave is propagating. Normally n also is a function of the light frequency. For a vacuum, we obviously have $n \equiv 1$. With these definitions, Equation 3.2 can also be written as

$$c_m = \frac{\Delta x}{\Delta t} = \nu\lambda = c/n. \tag{3.3}$$

Using Equation 3.3, one can calculate λ if ν is known, and vice versa. If the wavelength is measured in vacuum and if the international (SI) length unit *meter* is used, the conversion factor is defined as exactly

$$c = 299,792,458 \, m/s.$$

However, in astronomy, wavelengths are often measured in air. In this case, the conversion factor c_m depends (through n) on the air temperature, the atmospheric pressure, and the atmospheric composition, and a conversion of λ to ν requires an empirical derivation of c_m or n. In practice, this difficulty is avoided by calibrating wavelength scales with standard spectral lines of well-known wavelengths.

3.1.2 The General Case

So far, we have discussed a special case of light waves only. Many of the basic effects of optics can be well understood on the basis of this special case. For a realistic modeling of the propagation of light, however, a more general mathematical treatment of electromagnetic waves is needed. A discussion of the full theory is outside the scope of this introduction. However, for orientation, the following paragraphs briefly list the main modifications to the theory, which must be made if the simplifying assumptions made earlier are dropped, and when more general types of waves are considered:

- Nonplane waves: In the general case, the full electric field vectors must be used to describe the wave, and the amplitudes become functions of the space coordinates.
- Nonmonochromatic (multiple frequency) waves: In general, electromagnetic waves can be represented by a superposition of monochromatic waves. If the wave travels in a medium with a linear response (which is normally the case for optical wavelengths, but not necessarily at radio frequencies), the different frequencies can be treated separately. Otherwise their interactions must be included.
- Finite waves: The wave described by Equation 3.1 has neither a beginning nor an end. A realistic, finite wave can be described as a superposition of waves with different frequencies.
- Phase shift $\varphi \neq 0$: Phase angles become important if, within a light beam, phase differences are produced by variations of the refractive index, by path length differences, or by reflections. The phase angle is also needed for describing polarized light, in which the different space components of the electric vector are coupled by a constant phase relation.
- Conductive and dissipative media: Highly conductive media (such as metals) often (but not always) result in a light reflection at their surface. If light is able to enter a conducting medium, the waves interact with the local electrons, which results in an absorption and damping of the wave. In an absorbing

medium, the amplitude of a plane wave decreases exponentially with the path length according to

$$E(x) = E_0 e^{-\kappa x}, \tag{3.4}$$

where κ is called the *absorption coefficient*.

Although the electric and magnetic field strengths are always real functions, for mathematical derivations it is sometimes advantageous to describe light waves by complex exponential functions instead of the trigonometric function used in Equation 3.1. For this purpose, one makes use of the fact that for any independent variable x, we have

$$e^{ix} = \cos x + i \sin x. \tag{3.5}$$

Thus, we can describe a plane light wave also by

$$E(x, t) = E_0 e^{i(kx - \omega t)}, \tag{3.6}$$

where E and k can be complex functions. If $k = 2\pi/\lambda = \omega n c^{-1}$, the real part of Equation 3.6 is equivalent to Equation 3.1. However, Equation 3.6 also provides a convenient way to describe a damped wave. For this purpose, one defines a complex wave number $k_c = \omega c^{-1}(n + i n\kappa)$, where $n + i n\kappa$ is called the *complex refractive index*. With this definition, Equation 3.6 can be written as

$$E(x, t) = E_0 \exp\left[i\omega(c^{-1}n x - t)\right] \exp\left(-c^{-1}\omega n \kappa x\right) \tag{3.7}$$

or

$$E(x, t) = E_0 \exp\left[i(kx - \omega t)\right] \exp\left(-c^{-1}\omega n \kappa x\right). \tag{3.8}$$

The second factor of $E(x, t)$ obviously describes the exponential damping of the wave.

For harmonic oscillations, the energy is known to be proportional to the square of the oscillation amplitude. Thus, light intensities are given by the square of the amplitudes of the light waves. If the wave is described in complex form according to Equation 3.6, the square must be replaced by a product with the corresponding complex conjugate function. Because this product is always real, intensities are automatically real functions.

3.2 Measuring Frequencies

There are several ways to describe spectra. The observed flux can be presented either as a function of the frequency ν, as a function of the wavelength λ, or as a function of the wave number k. In general, using the frequency is preferable, as ν is an intrinsic property of the radiation, whereas the wavelength depends on the medium in which the wave is propagating. Being defined as the number of oscillations per unit time, ν can (at least in principle) be determined by counting the number of electric or magnetic field maxima, which occur during an accurately measured time interval. Counting oscillations is relatively easy in the radio bands up to 100 GHz. All that is needed for this purpose is an electronic counter and an accurate clock. Using frequency chains or laser frequency combs, frequencies can also be measured for IR, visual, and UV light waves. As explained in Chapter 6, at optical wavelengths these methods are used to calibrate spectroscopic observations. However, in the optical range, direct frequency measurements are not applied to astronomical light sources, because in these bands it is (so far) more convenient to measure wavelengths.

At X-ray and gamma wavelengths, the frequencies are often determined indirectly by measuring the energy of the observed photons. Since Einstein's fundamental 1905 work, we know that light is emitted in the form of light quanta or photons, which have individual energies of

$$E_{phot} = h\,\nu, \tag{3.9}$$

where $h = 6.62606896 \times 10^{-34}$ J s is Planck's constant. In contrast to the vacuum velocity of light, Planck's constant is a measured quantity, which is known only within certain error limits. Frequencies derived by means of Equation 3.9 obviously cannot be more accurate than h. However, the present relative error of h is $< 10^{-7}$. For astronomical applications this is always sufficiently accurate, as the direct measurements of photon energies have much larger errors.

As pointed out in Chapter 1, $h\nu$ can be measured if a photon is completely absorbed in a detector, and if the detector signal is proportional (or otherwise uniquely related) to the photon energy. Examples of such detectors are the X-ray CCDs. The resulting energy resolution depends on the photon energy and the detector technology. With current detectors, the energy resolution for soft X-rays is still relatively low. Therefore, at X-ray energies, photon-energy-sensitive detectors are used regularly for low-resolution spectroscopy, whereas for high-resolution X-ray spectroscopy, wavelength measurements give more accurate results.

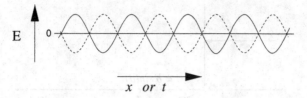

Figure 3.2. Two monochromatic plane waves with a $\lambda/2$ phase difference.

3.3 Measuring Wavelengths

3.3.1 Wavelength Measurements Using Interference

Important phenomena of all types of waves are the *interference* effects, which occur when waves are combined. Monochromatic waves with no phase difference simply add up, because the wave maxima and minima occur at the same time and position. However, as illustrated by Figure 3.2, if two waves of equal amplitude with a phase difference of $\lambda/2$ (or an angular phase difference of π) are combined, they exactly cancel each other. Between these extremes we get partial amplification or partial damping. If the amplitude is increased by the superposition of waves, we talk of *constructive interference*, and vice versa, an amplitude decrease is called *destructive interference*. These effects can be used to measure wavelengths.

The first one to determine the wavelength of light by means of an interference experiment was the English physician and scientist Thomas Young (1773–1829). Starting in 1801, Young investigated the diffraction of light at narrow slits. He first experimented with a single slit. He found that behind a narrow slit there is, in addition to light in the forward direction ($d = 0$ in Figure 3.3), light deflected to other angles d. Furthermore, he found that the observed light intensity varied periodically with the diffraction angle d, forming a "fringe pattern." This result can be explained, and the light distribution behind a slit can be calculated, using Huygens' principle, which postulates that every point that is reached by a light wave can be assumed to be the center of a new spherical wave propagating into the half-space in front of the original wave plane. Applying Huygens' principle to the slit experiment of Figure 3.3, we have to add up the new waves emerging from all points across the slit between A and B and propagating into the space to the right of the slit. The calculation (which is described in detail in optics textbooks) results in an intensity distribution behind the slit given by

$$I_s(d) = I_0 \frac{(\sin \delta_s)^2}{\delta_s^2},$$
(3.10)

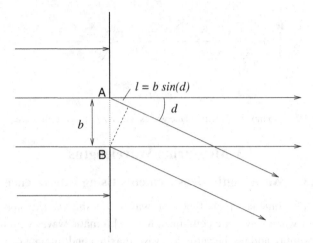

Figure 3.3. Diffraction of a plane wave at a single narrow slit in an opaque screen.
The slit is assumed to extend perpendicularly to the image plane, and it is assumed
to be much longer than the slit width b. The wave is assumed to propagate from
left to right.

where

$$\delta_s = \pi \lambda^{-1} b \sin d \qquad (3.11)$$

is, for a given diffraction angle d, just $1/2$ of the angular phase difference
$2\pi l/\lambda$ between waves originating from the two edges A and B of the slit
(where $l = b \sin d$; see Figure 3.3). Because the phase difference between A
and B is the maximum phase difference occurring for a given diffraction angle
d, δ_s can be regarded as the "typical" phase difference within a beam toward d.

As can be seen from Equation 3.10, the intensity I_s always has a maximum
at $d = 0$ where $\sin(\delta_s) \to \delta_s$. Moreover, from Equations 3.10 and 3.11 it is
clear that for a slit width $b \gg \lambda$ we have outside this maximum $\delta_s^2 \gg (\sin \delta_s)^2$,
which results in very low intensities for angles $d \neq 0$. Thus, for $b \gg \lambda$ we get
(as is to be expected) the result predicted by geometric optics. For a narrow slit
and monochromatic light, the periodicity of the sine function in Equation 3.10
results in an intensity distribution that varies periodically with d. These periodic
intensity variations are the "fringes" that were observed by Young.

According to Equation 3.10, the first minimum (or first "dark fringe") occurs
when $\sin d = \lambda/b$. Further minima are observed at $\sin d = m \lambda/b$, where m is
an integer. If b and d are measured, one can use Equation 3.10 to derive the
wavelength of the light. However, in practice, a single slit is not well suited
for this purpose, because outside the main maximum at $d = 0$ the intensity
is dominated by destructive interference. Therefore, the intensity of the first

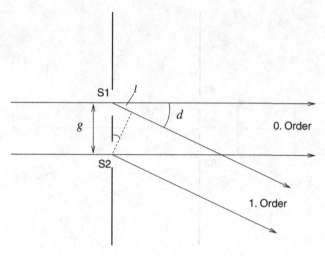

Figure 3.4. Thomas Young's double-slit experiment.

secondary maximum is only a few percent of the main maximum. This rapid intensity decrease makes it difficult to locate the first minimum accurately.

Much better results can be achieved by means of the interference pattern behind two closely spaced parallel slits. This "double-slit experiment" was pioneered by Young in 1802. It is schematically described in Figure 3.4. For this figure it is again assumed that a monochromatic plane wave enters from the left and reaches an opaque screen, whose plane is perpendicular to the propagation direction. Now, however, the screen contains two parallel slits, S1 and S2, with slit widths b. Because of the diffraction at the individual slits, light will again be deflected according to Equation 3.10 into a range of diffraction angles d. However, in addition to the interference of waves within each slit, we now also must consider the interference of the light from the two slits.

As shown in Figure 3.4, for a diffraction angle d the (mean) path difference of light from the two slits is $l = g \sin d$, where g is the distance between the two parallel slits. Obviously, we get constructive interference (or intensity maxima) if

$$l = g \sin d = m \lambda \tag{3.12}$$

(where the integer m is called the "order" of the maximum); we get destructive interference (or minima) if $l = (m + 1)\lambda/2$ and m is zero or even. Thus, the interference of the light from the two slits again results in a periodic fringe pattern, which is superimposed on the fringe pattern of the single slits.

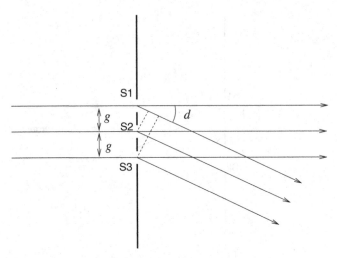

Figure 3.5. Light diffraction at three equidistant parallel slits.

By choosing a suitable slit distance $g > b$, one can ensure that one or several maxima (or orders) of the double-slit interference pattern are inside the forward maximum of the single slits. In this case, the corresponding angle can be easily measured, and the wavelength (in air) can be determined using Equation 3.12.

The accuracy can be improved further by adding more slits. Figure 3.5 shows the case of three slits. The distance between two adjacent slits is again assumed to be g. Thus, we still get constructive interference of the light from all adjacent slits if $l = g \sin d = m \lambda$. At the corresponding angles d, the path difference between the light from S1 and S3 is also a multiple of λ. Thus, interference between S1 and S3 reinforces these maxima. On the other hand, the beams from S1 and S3 reach a path difference of λ already when the path difference between two adjacent slits is $g \sin d = m \lambda/2$. However, in this case only the light from the slit combination S1 and S3 interferes constructively, whereas the other combinations result in destructive interference. Therefore, these secondary maxima are weaker than the maxima defined by Equation 3.12.

The presence of three slits also results in additional minima. Obviously, the interference between S1 and S3 is destructive when the light path difference is $\lambda/2$, which corresponds to path differences l between adjacent slits, which are multiples of $\lambda/4$. The corresponding angles d are between the main maxima and the minima of the two-slit fringe pattern. As a result, with three slits the maxima at $l = m \lambda$ are narrower.

3.3.2 Diffraction Gratings

The results for two and three slits can be easily extended to arrangements of larger numbers of N equidistant slits. Such an arrangement is called a *diffraction grating*. More specifically, because other types of diffraction gratings exist, it is called a *transmission diffraction grating*. The distance g between two adjacent slits is called the *grating period* or *grating constant*.[1] Independent of N, Equation 3.12 always defines the path length differences and the diffraction angles of the dominant maxima of the intensity distribution behind a transmission grating. With $N > 3$, there will be additional secondary maxima. However (for the reason given for the case $N = 3$), relative to the main maxima, with increasing N the intensities of the secondary maxima become progressively weaker. Moreover, owing to additional closely spaced minima, the main maxima will become progressively narrower. The corresponding intensity distribution $I(d)$ behind a grating can be calculated by adding up the interference of the contributions of all individual slits to the beam toward the direction d; that is, all contributions

$$E_p = E_0 \cos{[kx - \omega t + (p - 1)2\pi\, l\, \lambda^{-1}]}, \tag{3.13}$$

where $l = g \sin d$, and $p = 1, \ldots, N$. To calculate the sum, it is convenient to use the complex wave notation, where the sum in the argument of an exponential function can be converted into a product of exponentials. The result of the calculation (which is described in detail in optics textbooks) gives, for the intensity distribution behind a transmission grating with N slits,

$$I(d) = I_s(d)\frac{(\sin N\delta_g)^2}{(\sin \delta_g)^2} \tag{3.14}$$

where $I_s(d)$ is given by Equation 3.10 and

$$\delta_g = \pi l\lambda^{-1}. \tag{3.15}$$

The first factor in Equation 3.14 obviously results from the diffraction at the individual slits, whereas the second factor describes the effect of the interference of the light from the N slits. Equation 3.14 confirms that $I(d)$ has maxima at $l = m\lambda$. For $l \to m\lambda$ we have $\sin \delta_g \to \delta_g$ and $\sin(N\delta_g) \to N\delta_g$, and, consequently, $I(d) \to I_s N^2$. Outside these maxima the quotient of the squared sine functions has values that are not much larger than 1. Thus the intensity

---

[1] Although this notation is used in most of the physical and astronomical literature, in some publications the term *grating constant* is defined as g^{-1}. Because of this ambiguity, only the term *grating period* will be used in this book.

of the main maxima surpasses the intensity at other angles d by factors of the order N^2.

Equation 3.14 also shows that between the maxima at $l = m\lambda$ we have minima when l is an integer multiple of $N^{-1}\lambda$. Thus, at each maximum the nearest minimum corresponds to a light path difference

$$\Delta l = \pm N^{-1}\lambda. \tag{3.16}$$

If we plot $I(d)$ as a function of $l = g\sin d$, we get intensity maxima at $l = m\lambda$ with a width of about $2\Delta l$ and (for large N) very little intensity between these maxima.

So far we assumed that the incident light is monochromatic with a single wavelength λ. If we add light with a wavelength $\lambda + \Delta\lambda$, we get for this second wavelength a qualitatively similar fringe pattern, but with maxima at different values of d and l. If the difference between λ and $\lambda + \Delta\lambda$ is small ($\Delta\lambda \ll \lambda$), we can expect that at a given order the maxima of the two patterns can just be separated if their relative displacement corresponds to the relative path length difference described by Equation 3.16. In this case, we get

$$\frac{\Delta\lambda}{\lambda} = \frac{\Delta l}{l} = \frac{N^{-1}\lambda}{m\lambda} = (mN)^{-1}. \tag{3.17}$$

In spectroscopy, the *resolving power* of an instrument or the *spectral resolution* of a spectrum is usually expressed by the quantity $R = \frac{\lambda}{\Delta\lambda}$, which is the inverse of the ratio given in Equation 3.17. Thus, for the spectral resolution we get

$$R = \frac{\lambda}{\Delta\lambda} = m N. \tag{3.18}$$

Obviously, R depends only on the number of interfering beams and on the order of the interference.[2]

Although Equation 3.18 has been deduced in the context of diffraction gratings, we have not made use of the particular geometry of gratings. For the derivation of Equation 3.18 we assumed only an interference of N beams in the mth order, and that the wavelength is determined by measuring the path difference l. Therefore, Equation 3.18 is also applicable to other devices that measure spectra by inference.

In astronomical spectroscopy, it is usually not possible to measure $\Delta\lambda$ exactly as defined in the preceding paragraph. Therefore, in practice, different measures of the width of the image of a monochromatic line are used. Most

---

[2] In part of the astronomical literature, $\Delta\lambda$ (instead of the ratio R) is called the spectral resolution. Because the corresponding quantities have different dimensions, there is normally no danger of confusion. Nevertheless, if measured spectral resolutions are reported, it is advisable to note the definition and the method that were used to derive the corresponding values.

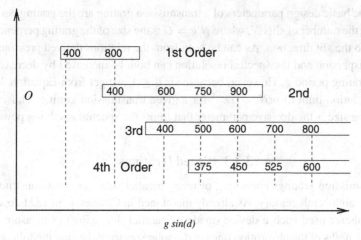

Figure 3.6. Wavelength overlap of the first four orders of a diffraction grating. The boxes indicate the typical wavelength range of a silicon-based CCD detector (350–1,000 nm). For orientation, some wavelengths (in nm) are indicated for the individual orders.

often, the full width at half the maximal intensity (FWHM) of an unresolved spectral line is taken to characterize $\Delta\lambda$. Moreover, deviations from the ideal optical geometry assumed earlier, optical aberrations, and other instrumental effects tend to result in an additional broadening of the spectral line images (see Chapter 4). Therefore, the effective resolving power of a real spectrometer is often significantly lower than that given by Equation 3.18. However, the product $m\,N$ is always a safe upper limit for R.

For the practical use of gratings in spectrometers, it is important to know, besides the spectral resolution R, the angular wavelength dispersion $\frac{dd}{d\lambda}$. By differentiating Equation 3.12, we get

$$\frac{dd}{d\lambda} = \frac{m}{g\cos d} = \frac{\sin d}{\lambda\cos d}. \tag{3.19}$$

Thus, for a given wavelength, the angular dispersion depends on the diffraction angle only.

A sometimes unwelcome property of diffraction gratings is the overlap of orders. According to Equation 3.12, the first-order maximum of a wavelength λ occurs at the same angle d as the second-order maximum of the wavelength $\lambda/2$. As illustrated in Figure 3.6, with increasing m, the wavelength separation becomes even smaller. Therefore, (as will be discussed in detail in Chapter 4) during the design of grating spectrometers measures must be taken to separate the orders or to suppress unwanted orders.

The basic design parameters of a transmission grating are the grating period g and the number of slits N, where $N\,g = G$ is the size of the grating perpendicular to the slit direction. As can be seen from the relations derived previously, the dispersion and the spectral resolution can both be increased by decreasing the grating period g. However, because $\sin d \le 1$, we get from Equation 3.12 a resolution limit of $mN < G\lambda^{-1}$ for a given transmission grating. Thus, the grating size is the decisive parameter that limits the spectral resolving power.

3.3.3 Blazed Gratings

Transmission gratings consisting of many parallel slits were first constructed in the eighteenth century. As already mentioned in Chapter 1, in 1821 Joseph Fraunhofer used such a device (made of parallel thin wires) to measure the wavelengths of the absorption lines in the solar spectrum. During the following decades, improved transmission gratings were produced by generating periodic opaque stripes on the surface of glass plates. These early gratings produced good spectra. However, according to Equations 3.14 and 3.10, in such simple transmission gratings most of the light ends up in the undispersed 0th order. Moreover, at the higher orders the light is split into two equal parts with different signs of d. As a result, these early gratings were rather inefficient, and their use in astronomy was initially restricted to solar observations. For the low light levels that are typical for stellar spectroscopy, gratings became of interest only after techniques had been invented that made it possible to concentrate most of the light into a single diffraction order $m \neq 0$. Such gratings are called *blazed gratings*. This type of diffraction grating soon became the most important spectral dispersion device in astronomy; today, blazed gratings are used at all wavelengths except for the gamma range.

The *blaze effect* was first achieved in reflection gratings. Blazed reflection gratings can be produced by cutting long grooves with suitable profiles into a plane surface. The grooves are made reflective by applying a thin metallic coating. Light beams reflected from the individual grooves interfere in the same way as the light from the slits of a transmission grating. They therefore produce the same type of interference patterns. Because the individual grooves act as narrow mirrors, in first approximation the light is deflected following the reflection law. However, owing to the diffraction of light at the narrow groves, the reflected light covers a range of angles around the direction of reflection. The width of the deflected beam again corresponds to the diffraction pattern of a single slit (given by Equation 3.10). In this way, we get a grating diffraction pattern that has its maximal intensity in the direction of the reflected light. By inclining the reflecting surfaces of the individual grooves relative to the

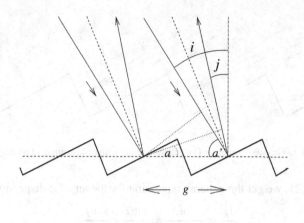

Figure 3.7. Cross section of the grooves of a blazed reflection grating. The grooves extend perpendicular to the image plane.

grating plane, it is possible to deflect most of the light into one preselected order. The angle between the normal to the reflecting groove surfaces and the normal to the grating plane is called the *blaze angle*. Light reaching the grating at the blaze angle is reflected back in the same direction. For this light, a reflection grating reaches its maximal efficiency. Schematically, the functioning of a blazed reflection grating is outlined in Figure 3.7. In this figure, a typical groove cross section and examples of incident and diffracted rays are plotted.

As in the case of the double slit and the transmission gratings discussed earlier, maxima of the diffracted light occur when the path length difference of light between adjacent grooves is $l = m\,\lambda$. However, at a blazed reflection grating, the incident light may be inclined to the grating normal (as in Figure 3.7). Therefore, the path length difference l now depends on the diffraction angle j as well as on the inclination angle i of the incident beam. From Figure 3.7 we get for l

$$l = g \cos a' + g \sin a = g \sin i + g \sin j, \qquad (3.20)$$

where we have made use of $a = j$ and $a' = \pi/2 - i$. Thus, the condition for the main maxima of the interference pattern becomes

$$l = g(\sin i + \sin j) = m\,\lambda. \qquad (3.21)$$

Equation 3.21 is called the *grating equation*. It is valid for all types of diffraction gratings. Equation 3.12 follows from Equation 3.21 for $i = 0$. By differentiating

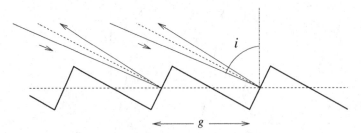

Figure 3.8. Groove geometry of an echelle grating with a blaze angle of 64 degrees.

Equation 3.21, we get the general expression for the angular dispersion

$$\frac{\mathrm{d}j}{\mathrm{d}\lambda} = \frac{m}{g\cos j} = \frac{\sin i + \sin j}{\lambda\cos j}. \tag{3.22}$$

The expression for the spectral resolution $R = m\,N$ remains valid, but, making use of Equation 3.21, R can now also be written as

$$R = \frac{\lambda}{\Delta\lambda} = m\,N = \frac{G}{\lambda}(\sin i + \sin j), \tag{3.23}$$

where

$$G = N\,g \tag{3.24}$$

is the total width of the grating. At present, gratings with dimensions up to about one meter are commercially available. Thus, for visual light ($\lambda \approx 500$ nm) and high angles i and j, resolutions up to $R \approx 4 \times 10^6$ are technically feasible with such gratings.

According to Equation 3.23, for a given grating size a maximal resolution is achieved if i and j approach $\pi/2$. Thus, for high-resolution spectroscopy, large values of i and j are of advantage. The corresponding groove geometry is illustrated in Figure 3.8. Because a large value of j corresponds to high order m, overlapping of orders is severe for such gratings. Their use therefore requires an efficient separation of orders. Normally this is achieved using a second dispersing device (e.g., a second grating or a prism) that produces a low-resolution spectrum perpendicular to the main dispersion. Because the resulting two-dimensional diffraction pattern (see Figure 4.16) resembles the rungs of a ladder, such spectra are called *echelle spectra* ("échelle" being the French word for ladder), and the corresponding high-order gratings are called *echelle gratings*.

From Figure 3.8 it is clear that an increase of the inclination angle j is limited by shadowing effects at the grating grooves, which reduce the efficiency at high values of i and j. As a compromise, echelle gratings are often operated with blaze angles of 63.5 degrees (corresponding to $\sin i = \sin j = 0.9$, $\tan i = 2$), where efficiencies of about 75 percent are still possible. Such gratings are called *R2 gratings*. However, echelle gratings with blaze angles of 76 degrees ($\tan i = 4$, R4) and 78.7 degrees ($\tan i = 5$, R5) are also commercially available. These gratings give higher resolutions, but their efficiencies are lower.

Reflection gratings can be made by cutting or burnishing the grooves directly into the grating substrate using a diamond tool. However, most present-day diffraction gratings are manufactured using a replication technique. For this purpose, a negative of the intended surface is produced by mechanical ruling or another suitable technique (discussed later in this section). These negatives are called *master gratings*, or simply "masters." The masters are then used as molds to cast replicas. Because it is very important for the shape and the spacing of the grooves to be constant over the whole grating surface, any wear of the ruling tool must be minimized. Therefore, mechanically produced master gratings are cut into relatively soft metal layers on top of a hard glass or glass-ceramic blank. Masters usually survive only a limited number of replications. Therefore, sometimes "submasters" are produced by the same replication process, and the final replicas are cast using submasters as a mold.

To produce grating replicas, a polished glass or metal plate is covered with a liquid resin. The master (or submaster) is pressed into the resin and left in contact, until the resin has partially hardened. A thin "parting layer" ensures that master and replica can be separated without damage to the delicate grooves. After the separation, the resin is fully cured.

The replication technique makes it possible to manufacture many gratings from a single ruled master. Although the production of a mechanically ruled master may take many months, the replication can be carried out on a time scale of days or weeks. Using a large master grating, replicas of any size smaller than the master can be produced. By replicating two or more very precisely aligned submasters on a common large substrate, it is even possible to produce replicas that are larger than the original master grating (see, e.g., Dekker et al., 1994, 2000).

In addition to mechanical ruling, several other methods can be used to produce master gratings. The most important alternative is photolithography. To obtain a master grating by this method, a photosensitive coating on top of a glass plate is exposed with laser light, which, by means of two-beam interferometry, produces a periodic illumination pattern. Chemical developing of the exposed coating then results in a periodic groove pattern in the coating. Different groove shapes can be achieved by means of inclination variations of the laser beams. As in the case of mechanically ruled masters, the actual gratings are replicated using a master as a mold. Gratings manufactured according to this process are called *holographically recorded diffraction gratings* (HRDGs) or simply *holographic gratings*.

Holographic gratings generally have the advantage of a very precise spacing of the grooves over the whole surface. However, the holographic process normally results in less-optimal groove shapes. Their blaze efficiency therefore tends to be lower than that of mechanically ruled gratings with similar parameters. Whereas mechanically ruled gratings may have peak efficiencies up to about 90 percent, holographic gratings typically reach only about 60 percent. On the other hand, mechanically ruled gratings often suffer from stray light, which is due to small ($<\lambda$) surface irregularities that are introduced by the burnishing process. Moreover, because the ruling of a large grating may take months, and because material fatigue and tool wear cannot be completely avoided, mechanically ruled gratings are not quite as regular as holographic gratings. Particularly critical are periodic irregularities, which can cause false diffraction orders, called *ghosts*. Because of the development of precise, interferometrically controlled ruling engines, today mechanically ruled gratings are much less affected by ghosts than in the past. However, holographic gratings are still of advantage when low stray light levels and a complete absence of ghosts are essential. Mechanically ruled gratings are to be preferred when high efficiency is the most important parameter.

Detailed information on the manufacturing, the properties, and the availability of mechanically ruled and holographic surface gratings can be found in the *Diffraction Grating Handbook* (Palmer and Loewen, 2005) and similar publications by the major suppliers of such devices.

The spectral dispersion effect of reflection gratings can be observed at all types of reflecting surfaces with a periodic relief. This includes natural objects (such as the feathers of birds) and technical devices such as compact discs and DVDs. This has prompted creative amateur astronomers to use sections of CDs for constructing simple and inexpensive astronomical spectrographs. For a recipe to build a CD-ROM spectrometer, see Köppen (2010).

3.3.4 Volume Phase Gratings

The gratings discussed so far all made use of periodic surface structures. However, diffraction gratings can also be based on three-dimensional periodic structures in the interior of a transparent optical element. An important example is the *volume phase gratings* (VPGs) (also called *volume phase(d) holographic gratings*, VPHGs). In its most simple version, a VPG consists of a transparent plane-parallel plate for which the refractive index n varies periodically in a direction perpendicular to the plate surfaces and in which light rays are

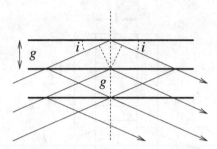

Figure 3.9. Bragg diffraction in a medium with periodically varying refractive index n. The thick lines indicate planes of constant refractive index. Light rays are indicated by thin lines. The refractive index is assumed to vary periodically with the spatial period g.

entering at an inclination to the planes of constant refractive index n, as indicated schematically in Figure 3.9. Variations of the refractive index along a light path result in reflection effects. Therefore, in the VPG, the light rays are partially reflected by the planes of higher-than-average refractive index n. The result of these reflections depends on the phase difference of the reflected rays. If the waves from all reflecting planes and from all points of a given plane have path length differences that are integer multiples of the wavelength, we get constructive interference and a strong intensity maximum. In other directions, we get destructive interference. From Figure 3.9, it can be seen that we get maxima when

$$2 g\, n_g \sin i = m\, \lambda, \tag{3.25}$$

where n_g is the refractive index inside the plate material and g is the spacing of the planes of constant refractive index. For the practical application of a grating, the corresponding relation with the inclination angle of the beam entering the surface of the device i_A is important. Assuming a refractive index $n_A \approx 1$ outside the grating and applying Snell's law, we get for this relation

$$m\, \lambda = 2 g\, n_g \sin i = 2 g\, n_A \sin i_A \approx 2 g\, \sin i_A. \tag{3.26}$$

Equation 3.26 is a special case of a *Bragg condition*. Bragg conditions, in general, describe the constraints on the light propagation, which result from the interference effects in three-dimensional periodic structures. This type of interference was first studied by W. L. Bragg and W. H. Bragg in 1913 in the context of the propagation of X-rays in crystals. The VPGs are a special application of the resulting theory.

Suitable transparent plane-parallel plates with an internal periodic modulation of the refractive index can again be produced by holographic (laser

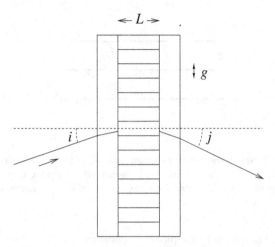

Figure 3.10. Schematic structure of a simple volume phase transmission grating. The layer with a periodically modulated refractive index is enclosed between two parallel glass plates. The planes of equal refractive index extend perpendicular to the image plane. Their spacing is g; L is the depth of the VPG.

interference) techniques. The interferometers that are employed for this purpose are similar to those used for holographic surface gratings. Approximately sinusoidal refractive index variations are normally generated in a chemically homogeneous material by modulating the density. Most VPGs use dichromatic gelatin, the density of which depends on the water content. Laser light can modify the water content locally. This effect can be used to produce density variations that result in refractive index variations up to about 0.1. Because dichromatic gelatin is hygroscopic, the active layer must be enclosed between glass plates and well sealed.

The layout and use of a simple VPG is indicated schematically in Figure 3.10. The interference of the waves that leave the grating follows the same rules as in the case of other gratings. The Bragg condition and the grating equation predict maxima at the same angles j. Hence, the grating equation (Equation 3.21) is also valid for VPGs. However, in addition to the grating equation, the light beams must meet the Bragg condition, that is, Equation 3.26. This can be used to obtain a strong blaze effect.

A property of Equation 3.26 is that it forms a strict relation between the entrance (and exit) angle of the light rays and the wavelengths. Thus, according to this equation, for a given inclination of the incident light beam, only one wavelength can pass through the VPG. This would be equivalent to the grating blazed very narrowly for just one wavelength. However, for Equation 3.9, it had been assumed implicitly that the planes of constant refractive index extend to infinity (i.e., $L \to \infty$ in Figure 3.10). For $L \to 0$, the Bragg condition disappears and we get a normal (unblazed) transmission grating. Intuitively, one might expect a VPG with a finite L to have properties between these extremes. The exact calculation of the interference in a VPG with a finite depth L shows that we indeed get a blaze effect for a finite bandwidth, which in first approximation is proportional to the ratio g/L. Unfortunately, the amplification factor also depends on L. If L becomes equal to or smaller than the wavelength, the amplification and the blaze effect disappear. Therefore, in the design of a VPG, the ratio g/L is critical. However, other parameters, such as the refractive index and its modulation amplitude, also enter into the blaze efficiency of VPGs. Optimizing the parameters therefore requires a detailed modeling.

The Bragg condition allows incidence angles $i = 0$. However, this results in a zero-order diffraction, which has no dispersion. Therefore, to use a VPG of the type indicated schematically in Figure 3.10, the incident beam must be inclined to the grating surface. By selecting a suitable tilt angle, one can obtain a blaze effect for several different diffraction orders and wavelength bands.

VPGs cannot be replicated, but are always originals. This results in longer production times and higher costs. However, VPGs have the advantage that each grating can be designed according to distinct specifications. The various free physical parameters of VPGs make it possible to produce gratings with a wide range of properties, although the interdependence of the parameters and physical constraints of the suitable optical materials set limits. Because Equation 3.10 is not valid for VPGs, strong blaze effects can be achieved for smaller grating periods g than is possible for surface relief gratings. With proper design, peak efficiencies near 100 percent can be reached. This makes it possible to produce low-order gratings with a relatively high spectral resolution. On the other hand, as explained earlier, VPGs with a high peak efficiency tend to have a narrow spectral range. Moreover, the need to meet the Bragg condition results in special constraints for the optical layout of spectrometers. Therefore, it is normally not possible to use VPGs in spectrometers that have been designed for other types of gratings. Useful relations for the design and the use of VPGs and an outline of their theory can be found in a concise review by Barden et al. (2000).

Because the surfaces of a VPG are normal glass plates, they are much more robust than those of surface gratings (which normally are destroyed if touched). Thus, VPGs have the operational advantage that their surfaces can be cleaned. This property of VPGs is particularly valuable for astronomical spectrographs, which are operated in dusty telescope buildings.

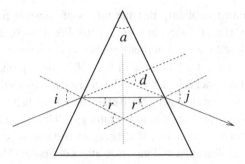

Figure 3.11. Refraction of monochromatic light in a prism.

3.3.5 Prisms

As noted in Chapter 1, during the first decades of astronomical spectroscopy prisms were the most popular dispersing devices. To separate different wavelengths, prisms make use of the wavelength dependence of the velocity of light in matter (as characterized by the refractive index n, defined in Equation 3.3). The light propagation in a prism is outlined in Figure 3.11. For this figure, a monochromatic plane wave is assumed to enter the prism from the left with an inclination i to the normal of the entrance surface. Outside the prism we assume (for simplicity) $n = 1$; for the prism material we assume $n > 1$. Snell's law predicts that the direction of propagation of the plane wave changes at the prism surfaces according to

$$\sin i = n \sin r \quad \text{and} \quad \sin j = n \sin r', \tag{3.27}$$

where the angles involved are defined in Figure 3.11. In this figure, the symmetric special case $i = j$ and $r = r'$ is plotted. Astronomical prism spectrometers normally try to stay as close as possible to this case, as the optical aberrations, which are caused by prisms, reach a minimum if $i = j$. However, because n depends on the wavelength, the conditions $r = r'$ and $i = j$ can be be achieved for only one wavelength, whereas for all other wavelengths we have $r \neq r'$, $i \neq j$.

For the symmetric case plotted in Figure 3.11 we get the relations $r = a/2$ and $i = d/2 + r = d/2 + a/2$, where d is the deflection angle. Inserting these relations into Equation 3.27, we get

$$n \sin (a/2) = n \sin(a/2 + d/2) \tag{3.28}$$

and

$$d = 2 \arcsin(n \sin a/2) - a. \tag{3.29}$$

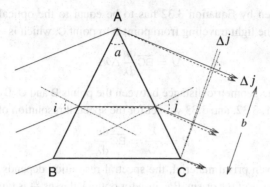

Figure 3.12. Two beams with a small wavelength difference $\Delta\lambda$ leaving a prism.

For small values of a and d, this can be approximated by

$$d \approx (n - 1)a. \tag{3.30}$$

Thus, in this approximation the deflection by the prism is proportional to its apex angle a. For a glass with a refractive index of $n = 1.5$, the deflection becomes about half the apex angle.

Although a prism separates wavelengths by refraction rather than by diffraction, diffraction effects limit its spectral resolution. Because prisms are zero-order devices, Equation 3.18 is not applicable for prisms. However, in first approximation the diffraction effects of a prism can be estimated from Equation 3.10 by inserting as "slit width" the beam width b, which is the projected width of the prism. Because for prisms we have $b \gg \lambda$, for a given wavelength λ the light will be diffracted into a narrow range of angles around j. According to Equation 3.10, the first diffraction minimum will occur at $j + \Delta j$ where $\sin \Delta j = \lambda/b$ or, as the angle is very small, with good approximation

$$\Delta j = \lambda/b. \tag{3.31}$$

For estimating the spectral resolution we assume that two spectral lines with wavelengths λ and $\lambda + \Delta\lambda$ can just be resolved if their refracted beams have an angular difference of Δj as given by Equation 3.31. From Figure 3.12 it can be seen that for light that passes through the apex (Point A) of the prism (practically outside of the prism material), Δj results in a geometrical (and optical) pathlength difference

$$\Delta l = b\Delta j \tag{3.32}$$

between the beams corresponding to λ and $\lambda + \Delta\lambda$, respectively. To produce an image, the wavefronts must be in phase for all rays. Thus, the pathlength difference given by Equation 3.32 has to be equal to the optical pathlength difference of the light traveling from point B to point C, which is

$$\Delta l = \overline{BC}\frac{dn}{d\lambda}\Delta\lambda, \tag{3.33}$$

where $\overline{BC}$ is the geometric distance between the points B and C. By combining Equations 3.31, 3.32, and 3.33 we get for the spectral resolution of the prism

$$R = \frac{\lambda}{\Delta\lambda} = \overline{BC}\frac{dn}{d\lambda}. \tag{3.34}$$

Thus, for a given prism material, the spectral resolution depends only on the length of the basis of the prism. Because for optical glasses $\frac{dn}{d\lambda}$ is typically of the order 10^3 cm^{-1}, a prism with a 10-cm base length has a maximum resolution of the order $R = 10^4$. To match the resolution of the largest gratings (about 10^6), a prism would need an impossible base length of several meters. Therefore, gratings and other dispersers based on interference effects are much better suited for high-resolution spectroscopy. On the other hand, the fact that prisms produce no perceptible higher-order spectra can be an important advantage for certain applications.

As in the case of gratings, the resolution derived earlier should be regarded as an upper limit only, as optical aberrations and other effects tend to lower the effective values of R.

3.3.6 Grisms

Besides gratings and prisms, in astronomical spectroscopy *grating prisms*, or *grisms*, play a major role. Because grisms can be blazed in the forward direction, they are particularly useful for multiobject spectroscopy with focal reducers, where one can switch between direct imaging and spectroscopy by inserting (or removing) a grism. The optical principle of a simple grism is outlined in Figure 3.13. As shown by this figure, the grism consists of a grating of the type discussed in Section 3.3.3 on the surface of prism. In contrast to the case discussed in Section 3.3.3, however, this grating is used in transmission instead of reflection. It therefore has no reflective coating. In principle, the grating can be ruled directly into the surface of a glass prism. However, it is easier and less expensive to make use of the replication process described in Section 3.3.3, using a standard (or custom-made) prism as a grating substrate.

The interference of the light rays from the grism and the resulting interference pattern again follow from Equation 3.21. However, in the case depicted in Figure 3.13, the incident beam reaches the grating from inside the prism, for which we have a refractive index $n > 1$. Thus, the optical path length is n

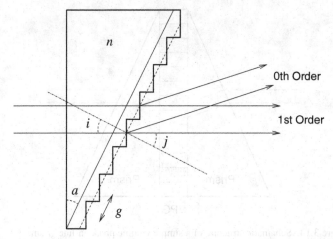

Figure 3.13. Light paths of a monochromatic beam through a grism that is blazed in the "forward" or "in-line" direction. The angles i and j are measured from the grating normal.

times the geometric path length. As a result, the grating equation for a grism of the type shown in Figure 3.13 becomes

$$g(n \sin i + \sin j) = m \lambda. \tag{3.35}$$

When using this equation, we must keep in mind that the angles i and j are on opposite sides of the grating normal. Thus i and j have opposite signs.

Figure 3.13 shows the important special case in which light passes the grating without deflection. Because this is the direction following from geometrical optics, we get a blaze effect for this angle. A grism with this property is called *forward blazed* or *in-line blazed*. In the configuration of Figure 3.13, the in-line blaze is achieved by making the entrance surface of the prism and the effective groove surfaces parallel and perpendicular to the incident beam. In this particular case, we have $i = -j = a$, where a is the apex angle of the prism. Consequently, in this case we get for the transmission maxima

$$m \lambda = g(n \sin a - \sin a) = g(n - 1) \sin a. \tag{3.36}$$

As in the case of blazed reflection gratings, the blaze efficiency increases with increasing (projected) grating period g. For $g \gg \lambda$, the efficiency can reach values close to unity, as the limit of geometric optics is approached. In this case, the intensity in the 0th order becomes negligible. As can be seen from Figure 3.13 and the grating equation, the deflection of the 0th order just corresponds to the refraction by the prism.

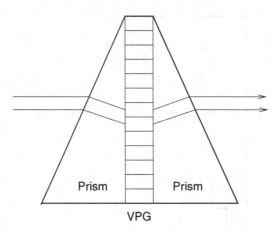

Figure 3.14. Schematic structure of a simple volume phase in-line grism.

As in the case of other gratings, the spectral resolution is proportional to $m\,N$. However, because the total phase shift over the grism is modified by the factor $n - 1$, for grisms of the type discussed previously we get $R = (n - 1)m\,N$. Most glasses have refractive indices of the order $n = 1.5$. Therefore, grisms with glass prisms usually have a somewhat lower resolution than reflection gratings with similar parameters. On the other hand, by using materials with refractive indices $n > 2$ – such as IR-transparent ZnSe ($n = 2.45$) or silicon ($n \approx 3.4$) – significantly higher resolutions can be achieved with grisms than with gratings.

For Figure 3.13, we implicitly assumed that the grating material and the prism have the same refractive index $n(\lambda)$. If the grating is ruled directly on the prism, this is, of course, automatically the case. If the grating is replicated, a transparent resin matching the refractive index of the prism must be used. Although there exist resins that match the refractive index (and its wavelength dependence) of some optical glasses (such as BK7), suitable resins matching highly refractive glasses and crystalline prism materials often do not exist. If the refractive indices do not match, refraction will take place at the interface between the prism and the grating material. If the refractive index difference is small, this can be partially compensated using suitably tilted groove surfaces. However, this tends to result in shadowing effects and reduced efficiency. Therefore, for prism materials with a high refractive index, a direct ruling on the prism is often the only feasible choice.

Instead of using surface relief gratings, grisms can also be made by combining prisms with volume phase gratings. An example of an in-line volume phase grism is schematically indicated in Figure 3.14. In this case, a VPG is sandwiched between two prisms. By choosing a suitable prism material and a

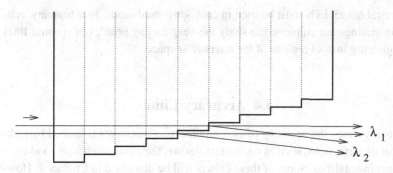

Figure 3.15. Schematic cross section and light rays through a transmission echelon grating.

suitable prism apex angle, the incident beam can be refracted into a direction inside the prism that meets the Bragg condition of the VPG. The outward beam is refracted back into the forward direction when leaving the second prism. Relative to surface grisms, volume phase grisms have the same advantages (and disadvantages) as VPGs do relative to conventional gratings. However, because the prism properties provide additional design parameters, volume phase grisms can be more easily adapted to conventional spectrograph designs. For a detailed discussion of the design and the use of volume phase grisms see Hill et al. (2003).

3.3.7 Echelon Gratings

Echelon gratings (EGs) are diffraction gratings with low line numbers N, which are operated at very high orders of typically $m \approx 10^4$ (as compared with typically 10^2 for echelle gratings). EGs combine some properties of grisms and echelle gratings. A cross section through a transmission echelon grating is given schematically in Figure 3.15. Originally, gratings of this type were produced by stacking identical glass plates with a thickness of about 1 cm in optical contact. A slight (of the order of millimeters) offset between the plate fronts resulted in a staircase-like surface, as indicated in Figure 3.15. EGs can be used in transmission (as indicated in Figure 3.15) or in reflection. In the latter case, a reflective coating is applied to their surface, and in Figure 3.15 the illumination would have to be from the right.

For many decades, stacking of highly polished glass or fused silica plates was the only method to produce EGs. Today, such devices can also be made by directly cutting step-shaped grooves into glass or aluminum blanks.

The echelon grating was invented by Michelson in 1898 for very-high-resolution ($R \approx 10^6$) laboratory spectroscopy. Because of their small free spectral range, EGs must be used in cross-dispersed mode. In astronomy, echelon gratings are employed to study the very narrow profiles of spectral lines originating in cool regions of the interstellar space.

3.4 Accuracy Limits

In many cases, the accuracy of astronomical measurements is limited by technical imperfections, such as instrument flexure, temperature changes, and electronic instabilities. Some of these effects will be discussed in Chapter 4. However, there are also accuracy limits that are more basic. Examples are imprecise or differing definitions of the units used and limits resulting from the quantum nature of light and energy.

3.4.1 Frequencies and Wavelengths

As noted before, the most basic spectral measurement is the derivation of a frequency – that is, counting the number of oscillations per unit time. The accuracy of such measurements is limited by the accuracy of the clock, which is used to measure time intervals, and by the precision of the unit of time. Accurate direct astronomical frequency measurements were first carried out in radio astronomy. Most radio astronomy was developed after the definition of the present-day official unit of time, which took place in 1967. This (IAU-sanctioned) time unit is the *international second* of the Système internationale d'unités (SI). It is defined by a spectral line of the cesium-133 atom. The SI second is realized by means of cesium clocks and other atomic clocks. Lines in the spectra of astronomical sources generally have a resolution that is much lower than the relative accuracy of these clocks. Therefore, transportable atomic clocks and/or standard frequencies emitted by global positioning satellites (also based on atomic clocks) provide a sufficiently accurate base for astronomical frequency measurements.

In contrast with laboratory spectroscopy, in astronomy the observer's reference frame is generally very different from that of the source. Thus, observed frequencies must be converted to the source reference frame or to one of the standard astronomical reference frames, such as the solar system barycenter. In the past, inaccuracies and errors in the reference frame conversions often introduced errors in published frequencies. Sophisticated modern computer codes avoid such errors, if they are properly applied.

As in the case of frequency measurements, contemporary astronomy (since 1960) uses SI units for wavelength measurements. The SI length unit is the *meter*. However, accurate wavelength measurements in the spectra of astronomical objects started in the nineteenth century, long before the current SI meter was defined in 1983. Sometimes historic measurements are still of interest (e.g., in the case of variable objects). Therefore, when using older spectroscopic data, one must take into account the complex history of the units that have been used for astronomical wavelength measurements.

In principle the use of the meter in spectroscopy began in 1868 when the Swedish astronomer Anders Jonas Ångström (1814–1874) started measuring wavelengths in the solar spectrum in units of 10^{-10} m. In a sense, this unit, denoted as the *Ångström* (often written as Angstrom, and abbreviated as either Å or A), is still in use today. In the nineteenth century, Ångström's unit became widely accepted, but with different definitions of the meter. An internationally agreed-upon definition of the meter was established only in 1875 by the international Meter Convention and the founding of the International Bureau of Weights and Measures (officially named *Bureau International des Poids et Mesures*, BIPM). However, spectroscopic measurements soon became more accurate than the accuracy of the BIPM meter definition. Moreover, the calibration of spectroscopes with the (mechanical) standard meter was difficult with the technologies of the nineteenth century. Therefore, in 1907 the International Union for the Co-operation in Solar Research (one of the first large international astronomical associations) redefined the Å unit by means of a standard spectral line (the 6438 Å line of cadmium). Although the new Å was (on the basis of interferometric measurements) defined to be as close as possible to 10^{-10} of the international meter, the more accurately defined new Å was now decoupled from the BIPM meter. Because of the higher accuracy and because spectrometers could be easily calibrated with the cadmium line, the new length unit for wavelength measurements was (with the consent of the BIPM) soon adopted by spectroscopists outside astronomy and even outside spectroscopy.

The two units, meter and Å, were used in parallel until 1960, when the official (BIPM) meter was redefined. The length of the meter was now (temporarily) also based on a spectral line, but on a different one than the Å. However, to retain the existing accurate wavelength tables (which already existed at that time), the new meter was fixed to be 10^{10} of the Å unit as defined in 1907. As a result, the meter experienced a small relative change (by about 2×10^{-7}), whereas the Å (now again $= 10^{-10}$ m) remained unchanged.[3] Hence, wavelength values expressed in Å and published since about 1907 correspond to the present (SI)

---

[3] The meter definition was changed again in 1983 (by giving a fixed value to the vacuum velocity of light in m/s), but this modification was much smaller and had no effect on the accuracy of astronomical spectroscopic data.

metric system, in spite of the 1960 (and 1983) changes in the meter definition (and length).

With the improved meter definition of 1960, the need for a special spectroscopic wavelength unit had, in principle, disappeared. Therefore, today the use of the unit $\text{Å} = 10^{-10}$ m, although still permitted, is officially discouraged by the BIPM. For this reason, in this book optical wavelengths usually are given in nanometers or microns. Exceptions are made in relations that are customarily expressed in Å, and in the context of some figures of optical spectra.

In the current astronomical literature, the Å is still used more often than the nanometer. One reason for the continuing popularity of the Å unit in astronomy is that lines in stellar spectra can usually be identified unambiguously by their wavelength in Å using a whole number.

If the unit is defined, the accuracy of wavelength data depends on the spectral resolution of the spectrometer, the resolution of the detector, the photometric accuracy of the flux measurements, the number of measurable spectral lines, and the accuracy of the calibration. Some of these issues will be discussed in the context of the reduction of spectroscopic data in Chapter 6. The most important application of accurate wavelength measurements are dynamical studies based on radial velocity variations. As noted earlier, grating spectrometers can give spectral resolutions $\lambda/\Delta\lambda$ of the order 10^6. Wavelength measurements involve the derivation of the positions of one or many line profiles. Because the profiles need not to be resolved for this purpose, wavelengths can be determined with accuracies that are significantly higher than R^{-1}. The best current radial velocity studies measure wavelength shifts that correspond to only a few times 10^{-9} of the observed wavelength (see, e.g., Mayor et al., 2003).

3.4.2 Photometric Accuracy

As noted in Section 3.2, light consists of photons with individual energies of $h\nu$, where h is Planck's constant and ν is the frequency of the radiation. Each astronomical spectral measurement records a certain amount of energy, which (assuming a linear detector) is given by

$$E_t = F_\nu\, A \Delta\nu\, \Delta t\, \epsilon, \tag{3.37}$$

where F_ν is the spectral flux from the observed object, A the light-collecting surface of the telescope, $\Delta\nu$ the frequency bandwidth, Δt the integration time, and ϵ the total instrumental efficiency. For monochromatic light, the total number of recorded photons then becomes

$$n_{phot} = \frac{E_t}{h\nu}. \tag{3.38}$$

Because in most astronomical light sources the photons are emitted as independent stochastic events, a measurement of n_{phot} (or E_t) has an unavoidable statistical uncertainty (and statistical mean error) of

$$\Delta n = \sqrt{n_{phot}}. \tag{3.39}$$

Δn is called the *photon noise*. In addition to the object flux, the detector also receives light from the sky background underlying the object. The photons from the background n_{sky} increase the photon noise (but not the signal n_{object}). Therefore, in first approximation, we get for photometric measurements a *signal-to-noise ratio* (S/N) of

$$S/N = \frac{n_{object}}{\sqrt{n_{object} + n_{sky}}}. \tag{3.40}$$

In actual observations, there are always additional noise sources (owing to the detectors and other instrument components) that result in additional terms in the denominator of Equation 3.40. Thus, in practice, Equation 3.40 provides an upper limit for the achievable S/N only. More accurate expressions for S/N have been derived for the different types of detectors and applications. Corresponding formulae can be found in textbooks on photometry and in the instrument handbooks on the Internet pages of the major observatories.

The order of magnitude of the photon flux F_p from an astronomical object (in photons per second, per m², and per Å wavelength bandwidth) can be estimated from the simple relation

$$\frac{N_{phot}}{A \Delta t} = F_p = 10^{7-0.4 m_V} \, \text{m}^{-2} \text{Å}^{-1} \text{s}^{-1}, \tag{3.41}$$

where m_V is the object's magnitude in the visual (Bessell V) band. As examples, we note that for a bright star with $m_V = 10$ (corresponding to a spectral flux of about 0.36 Jy) observed with an 8-m telescope, we can expect about 5×10^4 photons $\text{s}^{-1} \text{Å}^{-1}$, whereas a distant galaxy with typically $m_V = 25$ (corresponding to a spectral flux of about 0.4 μJy) results in about 0.05 photons $\text{s}^{-1} \text{Å}^{-1}$.

As shown by Equation 3.38, for a given flux or recorded energy the number of recorded photons decreases inversely proportionally to the frequency or photon energy. Therefore, photon noise is particularly obstructive to X-ray and gamma astronomy. To compensate for the scarcity of photons, at high gamma energies very large effective telescope areas are required. At IR and radio frequencies the number of photons is larger and photon noise is less important. However, Equation 3.40 is applicable only if individual photons can be recorded, which requires that the energy of the individual photons be larger than the thermal energy fluctuations of the detector. Assuming thermal equilibrium conditions, this results in the condition

$$h\nu > kT, \tag{3.42}$$

where k is the Boltzmann constant and T is the temperature (in Kelvin). At normal atmospheric temperatures, $h\nu = kT$ is reached at a wavelength of about 50 μm. By cooling detectors with liquid helium, the photon detection limit can be extended to the mm-wave range. If $h\nu < kT$, the thermal fluctuations (i.e., kT) limit the detection of electromagnetic radiation.

Whereas the effects discussed in this subsection primarily determine the accuracy of spectral flux measurements, they also affect wavelength derivations. Normally wavelengths are measured (directly or indirectly) by matching model or template line profiles to observed line profiles. Noisy observed profiles therefore tend to result in an uncertain match and inaccurate wavelengths. Particularly large errors are expected to occur when weak or blended absorption lines are distorted by photon noise. The quantitative effects of photometric errors on wavelength measurements depend on the observed spectra, as well as on instrumental properties. For a detailed discussion of this topic see, e.g., Hatzes and Cochran (1992).

4

Optical-Range Grating and Prism Spectrometers

This chapter describes the spectroscopic techniques that are used in most near-UV, visual, and IR spectrometers. Spectroscopy in this range is particularly important for astrophysics, and instruments for these wavelengths can be constructed using conventional optical elements, such as lenses, prisms, gratings, and normal-incidence mirrors. These instruments share many common features.

4.1 Commercially Available Spectrometers

Because optical spectroscopy plays an important role in many different branches of science, medicine, and industry, many manufacturers of optical instrumentation offer commercially produced spectrometers of different types. Most of these devices are not suited for the low light levels from astronomical sources. However, in addition to some instruments that have been designed specifically for astronomical applications (mainly for amateurs), there are commercial "low light level" general-purpose spectrometers on the market, which can be (and are) operated successfully for low-resolution spectroscopy at small telescopes. These commercially available instruments have the advantage of being complete systems, which include high-quality CCD or IR detectors. In many cases they can be conveniently connected to the USB port of a computer for instrument control and data readout. Usually the light from a telescope must be fed to the spectrometer using an optical fiber (see Section 4.5), but some of these spectrometers can be attached directly to a telescope focus. Observers interested in commercially available spectrometers will have no difficulties finding the addresses of potential suppliers by searching for optical spectrometers in the Internet.

81

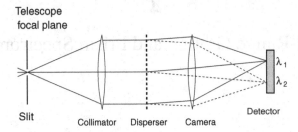

Figure 4.1. Basic optical layout of a simple slit spectrometer.

4.2 Basic Components of Astronomical Spectrometers

The low-light-level spectrometers offered by industry are often adequate for low-resolution spectroscopy of single objects at small telescopes. However, for reasons that will soon become clear, they cannot be adapted to large astronomical telescopes and for efficient multiple-object work. Therefore, most current astronomical spectrometers are custom-designed instruments that have been built specifically for astronomical research. Often the special requirements of astronomy result in technical concepts and solutions that are not found in other applications. Some of these specific concepts are discussed here. For the following sections it is assumed that the reader has some basic understanding of optical technologies. Introductions to this field can be found in all textbooks on optics and of optical engineering, such as those by Born and Wolf (1987) and Malacara and Thompson (2001). A comprehensive overview of present-day optical technologies is given in the four volumes of the *Handbook of Optics* (Bass, 1995). Valuable information on new developments in optics can be found in tutorial articles published regularly by the Society of Photo-Optical Instrumentation Engineers (SPIE). Part of these tutorial publications are publicly available through the SPIE Web site (http://spie.org). The SPIE also regularly organizes meetings on new developments in astronomical optics, and many current (and future) astronomical spectrometers have been first described in SPIE publications.

As noted in Chapter 1, since the late nineteenth century most astronomical spectrometers followed the design of Herman Carl Vogel, which in turn was based on the laboratory spectroscope of Kirchhoff and Bunsen (see Figures 1.4 and 1.3). Schematically, this design is outlined in Figure 4.1. Light entering a telescope is focused onto a spectrograph slit in the telescope's focal plane. The divergent beam behind the slit is made parallel by a collimator. This parallel beam (now consisting of plane waves) is split into a spectrum by a dispersing optical element. The dispersed beam consists of a continuum of parallel beams with different wavelengths and different directions. The camera focuses these chromatic beams to different positions on the detector.

As in most schematic drawings in this book, in Figure 4.1 the collimator and camera optics are schematically indicated as convex lenses. In real spectrometers, different optical components of the same functionality (such as concave mirrors, or complex multicomponent optical systems) may be used. The disperser can be a grating, a prism, or a grism.

Because the slit is located in the focal plane of the telescope, an image of the sky is produced on the screen that contains the slit. Normally the surface of the slit plane facing the telescope is polished and coated to act as a mirror. In this case, the sky surrounding the slit can be observed with a microscope or a TV camera. The resulting "slit images" can be used to position the target into the slit center and (using reference stars in the field) to control the tracking of the telescope during the spectroscopic exposure.

The optical system outlined in Figure 4.1 has various free parameters, such as the properties of the disperser, the focal length of the collimator, and the focal length of the camera. However, these parameters are not completely unrelated. To avoid light (or resolution) losses, the collimated beam must match the size of the disperser. The camera focal length determines the scale and size of the spectrum on the detector. Thus, the camera parameters must match the detector properties. Finally, the camera aperture must be large enough to accept the dispersed beam.

As pointed out in Chapter 3, if a grating is used as the disperser, the diffraction orders must be separated, or unwanted diffraction orders must be suppressed. This can be achieved by inserting order-selection filters into the beam, or by means of a "cross-disperser," whose dispersion is perpendicular to that of the main disperser. Cross-dispersed (or echelle) spectrometers contain the basic elements of Figure 4.1, but they often differ in details. Therefore, cross-dispersed spectrometers will be discussed separately in Section 4.4.6.

An unwelcome property of the simple design outlined in Figure 4.1 is the widening of the dispersed beam between the disperser and the camera. Obviously, this effect is particularly pronounced at high spectral dispersion and when the distance between the disperser and the camera is large. In this case, very large camera apertures may be required. During the photographic era, this problem was solved by using Schmidt cameras with large spherical mirrors (see, e.g., Figure 1.5), and the spectra were recorded on special large photographic plates. However, the smaller size of linear detectors and present-day larger gratings easily result in requirements that are outside the technical feasibilities of cameras that image the dispersed beam directly. Therefore, modern high-resolution spectrometers often make use of the so-called *white pupil*

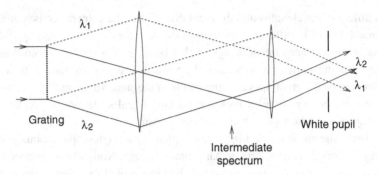

Figure 4.2. Schematic example of a white pupil configuration.

concept, which was introduced in 1972 by the French optician and astronomer André Baranne in the context of an early electronic detector system (Baranne and Duchesne, 1972). A schematic example of a white pupil configuration is given in Figure 4.2. As in Figure 4.1, the monochromatic parallel beams from the disperser are focused to produce a spectrum. However, in contrast to Figure 4.1, no detector is placed at the position of this (intermediate) spectrum. Instead, using an additional optical element (the *pupil optics*), the divergent monochromatic beams of the intermediate spectrum are converted into parallel beams, which intersect at some place behind the pupil optics. At this intersection of the monochromatic parallel beams we get a scaled image of the disperser. Because all monochromatic beams meet again in this plane, all colors are present over the surface of the corresponding pupil. A screen introduced into the beams at this position would therefore appear white. On the other hand, the full spectral information is present because the angles at which the light passes through the pupil depend on the wavelength, and a camera placed at the white pupil position records a spectrum.

Compared with the simpler design of Figure 4.1, the scheme of Figure 4.2 has the advantage that the size of the white pupil, and consequently the size of the camera aperture, can be adjusted by selecting a suitable focal length and position of the pupil optics. In the schematic figure, all optical elements are again indicated as convex lenses. In practice, reflective optical components are preferred for the pupil optics, because chromatic aberrations of lenses make it more difficult to realize a well-defined white pupil.

Apart from adapting large astronomical spectrographs to modern detectors, the white pupil concept is a valuable tool in the design of echelle spectrometers, in which a white pupil is the most favorable place for installing the cross disperser (see Section 4.4.6). Moreover, in all kinds of spectrometers, a white pupil can be used to efficiently reduce the straylight that is produced by gratings and other optical components.

4.3 Slitless Spectroscopy

In the simple spectrometer of Figure 4.1, the collimator is used to convert the convergent beam from the telescope objective (or mirror) to a parallel beam (or a plane wave). Many astronomical objects (such as stars, QSOs, and distant compact galaxies) are pointlike sources at large distances. Light waves reaching the telescope from these sources are, with very good approximation, already plane waves. Therefore, for such objects, in principle, the telescope, the slit, and the collimator of Figure 4.1 can be omitted and spectra can be taken by simply pointing the disperser-camera combination directly to the sky. In fact, during the early years of astronomical spectroscopy, many stellar spectra were taken according to this scheme. For this purpose, large, low-dispersion prisms or transmission gratings were placed in front of astronomical telescopes, which acted as large photographic cameras. Without the objective prism (or grating), the telescopes produced images of all stars in the field of view. With the prism in front of the objective, spectra were recorded for all the stars in the field. Because the prisms were mounted on top of the telescope objectives, this spectroscopic technique became known as *objective-prism spectroscopy*. Of course, the incident and the refracted beams of a prism have different directions. Therefore, for observations with an objective prism, the telescope must be pointed with a coordinate offset relative to direct imaging observations of the same field. If the prism parameters are known, this offset can be easily calculated and applied.

On objective-grating spectra, absolute wavelength measurements are possible using the zero-order images as zero points. Absolute wavelengths in objective-prism spectra can be derived making two exposures with different prism orientations. The orientations must differ by 180 degrees, and the coordinate offsets have to be modified correspondingly. This procedure results in two spectra (with different sign of the dispersion) on the same plate. Examples of such objective-prism spectra are presented in Figures 4.3 and 4.4.

Among the historic achievements of objective-prism spectroscopy were the first large catalogs of stellar spectra, such as the *Henry Draper Catalogue*, which was published in nine volumes of the *Annals of Harvard College Observatory* between 1918 and 1924. It listed the spectral classes of more than 200,000 stars. Later, the objective-prism technique was adapted to Schmidt telescopes, in which it was used extensively until the end of the photographic era.

Because prisms or gratings are limited to sizes of about 1 meter, at larger telescopes the objective prism technique cannot be applied directly. However, slitless spectra of point sources similar to those produced by the objective prism technique can be obtained using the basic arrangement of Figure 4.1 by simply

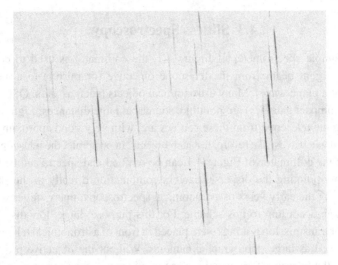

Figure 4.3. Section of a photographic objective-prism spectrogram. This plate was taken for radial velocity measurements. Therefore, for each object, two spectra with a different sign of dispersion are present to provide a reference for the wavelength scale. Image: Plate archive of the Heidelberg Observatory.

Figure 4.4. Another section of the photographic objective-prism spectrogram of Figure 4.3, showing some of the shortcomings of this technique. Because of the high density of stars in this field, there are overlapping spectra. Unresolved spectra of faint stars near the image center result in a nonuniform background. In the lower right corner, two spectra of an extended source are shown. Because of the extension in the dispersion direction, all spectral features are blurred in these two spectra.

removing the screen containing the slit from the focal plane. In this case, again, each point image in the focal plane results in a spectrum. Many modern low-dispersion spectrometers have this capability and offer this version of slitless spectroscopy.

Relative to single-object slit spectroscopy, the simultaneous recording of many spectra by means of slitless spectroscopy can obviously save much valuable observing time. However, this technique also has drawbacks. The most critical disadvantages are as follows:

- Sharp spectra are obtained for point sources only.
- Depending on the density of objects and the spectral dispersion, overlaps of the individual spectra will occur. In this case, partial or all spectral information is lost.
- Whereas in the case of slit spectra only the sky contribution within the slit is superposed on the object spectrum, in slitless spectra the whole field contributes to the sky background. Moreover, owing to unresolved spectra of faint objects, this background tends to be nonuniform.
- Seeing effects and telescope tracking errors directly influence the spectral resolution.

The higher sky background is most critical at large telescopes, which are designed for the spectroscopy of faint objects. Therefore, at large ground-based telescopes slitless spectroscopy today is limited to relatively few special programs. On the other hand, this technique is still applied successfully in space astronomy, for which the sky background is lower and the absence of seeing reduces the problem of overlapping spectra. Important examples of slitless spectrometers in space are the corresponding modes of the Hubble Space Telescope (see, e.g., Hartigan et al., 2004; Pasquali et al., 2006) and of the GALEX UV space observatory (see Chapter 7). Slitless spectroscopy will also be available on the future James Web Space Telescope. Another notable example of slitless spectroscopy is the spectrometer of the planned GAIA astrometric satellite (see http://gaia.esa.int).

4.4 Single-Object Slit Spectrometers

Because of the problems of slitless spectroscopy listed previously, most optical astronomical spectrometers are equipped with slits, or they are coupled to a telescope using optical fibers. In the latter case (which is outlined schematically in Figure 4.5 and discussed in detail in Section 4.5), the fiber exit takes over the function of the spectrograph slit.

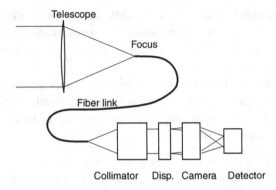

Figure 4.5. Schematic layout of a fiber-coupled single-object spectrometer.

If the length of a spectrograph slit (perpendicular to the dispersion) is larger than the spatial resolution element, spectra of distinct features along the slit can be recorded. Spectral images of this type are called *long-slit spectra* (even in cases in which only very few image elements can actually be resolved). Classical examples are spectra of inclined disk galaxies, where a single spectrum with a long slit covering a galaxy image can be used to derive the rotation law of the system.

To produce spectra of multiple objects simultaneously, modern spectrometers often contain several or many slits (or fibers). However, high-resolution spectrometers and dedicated long-slit instruments are still designed for single-object observations. Therefore, we shall start with a discussion of the simpler single-object spectrometers. Multiple-object spectrographs are basically extensions of these single-object instruments.

The basic principle of a single-object slit spectrometer was introduced in Figure 4.1. For simplicity, this figure showed the special case in which the incident collimated beam and the mean direction of the dispersed beam are both perpendicular to the plane of the disperser (corresponding to $i = -j$ in Figure 3.13). However, as pointed out in Section 3.3.3, most contemporary astronomical spectrometers use blazed reflection gratings in which i and j have the same sign and may have any value relative to the grating normal. How the collimated beam entering the grating and the beam entering the camera are related in this more general case can be seen from the schematic in Figure 4.6. In this figure, the angles i and j are defined as in Section 3.3.3 (i.e., i and j are always measured from the grating normal). As shown in Figure 4.6, in the general case the width of the incident collimated beam D and the width of the diffracted beam D' are different and related to the angles i and j according to

$$D = G \cos i \quad \text{and} \quad D' = G \cos j, \tag{4.1}$$

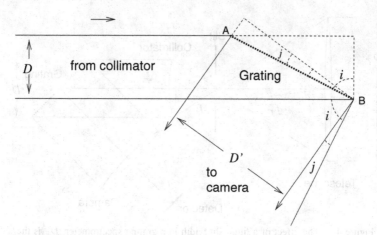

Figure 4.6. Incident and diffracted beam in a spectrometer that uses a blazed reflection grating.

where $G = N g$ is the width of the grating (see Equation 3.24). The ratio between the beam widths then is

$$\frac{D'}{D} = \frac{\cos j}{\cos i}. \tag{4.2}$$

These relations can be used to rewrite Equation 3.22 for the angular dispersion as

$$\frac{dj}{d\lambda} = \frac{m}{g \cos j} = \frac{\sin i + \sin j}{\lambda \cos j} = \frac{P}{D'\lambda}, \tag{4.3}$$

where

$$P = mN\lambda = m\frac{G}{g}\lambda \tag{4.4}$$

is the optical path length difference between the light rays from two opposite edges of the grating.

Of practical importance for the observer is the so-called *linear dispersion* (D_{linear}), which gives the scale of the spectrum on the detector. In linear approximation (expected to be valid near the center of the field of the spectrograph camera), this scale obviously is given by the product of the camera focal length f_{cam} and the angular dispersion. Although this product is sometimes used to characterize the scale on the detector, normally the linear dispersion is defined by the inverse of this product – that is, by

$$D_{linear} = (\frac{dj}{d\lambda} f_{cam})^{-1} = \frac{g \cos j}{m f_{cam}}. \tag{4.5}$$

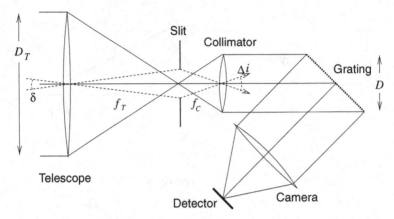

Figure 4.7. The effect of a finite slit width in a grating spectrometer. D_T is the telescope aperture; D is the diameter of the collimated beam; f_T and f_C are the focal lengths of the telescope and the collimator, respectively.

Although D_{linear}, as defined here, is dimensionless, it is usually expressed in units of Å/mm or nm/mm.

A related quantity is the *pixel scale*, which is defined as the wavelength interval corresponding to one detector pixel (or picture element). It is usually expressed as Å (or nm) per pixel.

4.4.1 Slit Width and Spectral Resolution

So far, we have assumed that the combination of slit and collimator optics results in a single parallel beam. This assumption is correct if the slit acts as a point source and if its width is negligible. In practice, however, spectrograph slits have a finite width. In order not to waste light, for small extended objects the slit is often opened to match the size of the object's image. For point sources, the slit width should at least match the diffraction pattern in the focal plane. If the image blurring by the turbulent Earth atmosphere (called *seeing* in astronomy) is not corrected, the slit width should match the extent of the seeing pattern, which even at good sites rarely has an FWHM smaller than about 0.5 arcsec.

In geometric approximation, the effect of a finite slit width is indicated schematically in Figure 4.7. For reasons that will become clear soon, in this figure, the telescope has been included in addition to the spectrometer optics. As can be seen from Figure 4.7, because of the finite slit width, light rays with incident angles i, which differ from that corresponding to the slit center, will reach the grating. The range of these angles i is indicated in Figure 4.7

by dashed lines. For small slit widths, the total range of these angles Δi is given by the ratio between the slit width and the focal length of the collimator. Some of the light with angles i in this range will be diffracted to the same angle j as the light from the center of the slit. However, according to the grating Equation 3.21, a different value of i (at constant j) corresponds to a different wavelength λ. Thus, the finite slit width results in an uncertainty of the wavelength diffracted toward an angle j. This effect obviously reduces the spectral resolution. Quantitatively, the wavelength uncertainty due to the finite slit width can be estimated from the product

$$\Delta\lambda = \Delta i \frac{d\lambda}{di}. \tag{4.6}$$

Like the spectral dispersion $\frac{dj}{d\lambda}$, the derivative $\frac{di}{d\lambda}$ (and its inverse) can be calculated by differentiating the grating Equation 3.21. However, for calculating the dispersion, i was assumed to be constant, whereas now j is constant and i is the variable. Because the grating equation is symmetric in i and j, the differentiation gives (analogous to Equation 4.3)

$$\frac{di}{d\lambda} = \frac{m}{g\cos i} = \frac{\sin i + \sin j}{\lambda\cos i} = \frac{P}{\lambda D}. \tag{4.7}$$

As shown by Figure 4.7, a given slit width s also corresponds to a distinct angle δ on the sky. This angle is called the *projected slit width*. For small angles, we obviously have

$$f_T\delta = s = f_C\Delta i, \tag{4.8}$$

where f_T and f_C are, respectively, the focal lengths of the telescope and of the collimator. According to Figure 4.7, these quantities are related to the apertures of the telescope and the collimator (D_T and D, respectively) by the equation

$$\frac{\Delta i}{\delta} = \frac{f_T}{f_C} = \frac{D_T}{D}. \tag{4.9}$$

By combining Equations 4.6, 4.7, 4.9, and 4.1, we get for the spectral resolution

$$R = \frac{\lambda}{\Delta\lambda} = \frac{\lambda}{\Delta i \frac{d\lambda}{di}} = \frac{P}{D\Delta i} = \frac{mN\lambda}{D_T\delta} = \frac{D}{D_T\delta} \frac{\sin i + \sin j}{\cos i}. \tag{4.10}$$

A comparison with Equations 3.18 and 3.21 shows that the result of Equation 4.10 is identical with that derived for the grating alone if

$$\delta = \frac{\lambda}{D_T}. \tag{4.11}$$

However, apart from a factor of 1.22, the term λ/D_T corresponds to the angular diameter the central maximum of the diffraction pattern of a circular telescope aperture (see, e.g., Kitchin, 2008). In the literature, this quantity is called the *diffraction limit* of the telescope. Thus, we get the important result that the spectral resolution approaches the value estimated already for the grating alone, if the projected spectrograph slit width is about equal to the telescope's diffraction limit.

Because of the finite angular size of many astronomical objects and because of the atmospheric seeing, often slit widths must be used that exceed the diffraction limit by a large factor. In this case, the spectral resolution may be well below the theoretical limit. If we observe at a wavelength $\lambda = 500$ nm with a seeing-matched projected slit width of one arcsec and a 10-m telescope, the spectral resolution R is only about 10^{-2} of the diffraction-limited case.

The preceding estimate shows that, for point-source observations with large telescopes, the spectral resolution can be improved significantly by reducing the slit width to values close to the diffraction limit. A smaller slit width also results in a lower sky background per spectral resolution element. The latter effect is particularly important for observations in the thermal IR, where the sky is always bright. Therefore, practically all modern large telescopes are equipped with *adaptive optics* (AO) systems that can eliminate, or at least reduce, the atmospheric seeing effects. Most existing AO systems were designed for IR applications, but AO techniques for the visual range are under development at various observatories. All these systems correct the wavefront deviations produced in the Earth's atmosphere by reflecting the incident light beam from a rapidly deformable mirror, which compensates for the atmospheric effects. For this purpose, a fraction of the beam is routed to a wavefront sensor that maps the incoming wavefronts observed from a reference source, such as an unresolved star. By comparing the observed wavefront with that expected for a distant point source, the required deformation of the adaptive mirror is computed and the mirror is deformed using piezoelectric or magnetic actuators. Depending on the site and the wavelength, the atmospheric wavefront fluctuations occur on timescales between 0.1 seconds and 10^{-3} seconds. Thus, the corrections must be applied with frequencies up to about one kHz. An example of the performance of a modern AO system is presented in Figure 4.8. The reader interested in more details of the AO technique can find a comprehensive description of the theory and the practice of AO systems in the textbook *Adaptive Optics in Astronomy* by F. Roddier (1999).

According to Equation 4.10, for a given projected slit width δ and given angles i and j, the spectral resolution is proportional to the ratio D/D_T. This is an important result, as it shows that for a given angular extent of the target at a larger telescope, a proportionally wider collimated beam and a correspondingly larger grating are required to reach the same spectral resolution. On the other hand, Equation 4.10 also shows that the spectral resolution is independent of

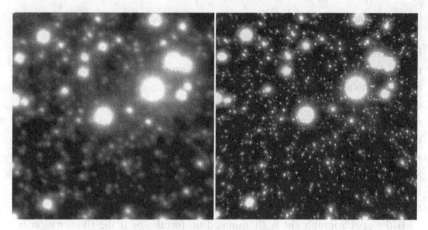

Figure 4.8. Example of seeing compensation by means of adaptive optics. Shown are two images of the globular cluster M13, taken with the 8-m Gemini-North telescope on Mauna Kea, Hawaii. The image on the left is a normal H-band (1.65 μm) image obtained under good seeing conditions with an angular resolution of about 0.26 arcsec. The image on the right was taken with the same telescope using an AO system. The resolution of that image is about 0.06 arcsec, which is close to the diffraction limit of the telescope at this wavelength. Images courtesy of Gemini Observatory/AURA.

the telescope size and is determined only by the properties of the spectrometer, if diffraction-limited observations of point sources are made.

Another important conclusion from Equation 4.10 concerns the spectroscopy of large extended objects with a uniform surface brightness and a uniform spectrum (such as large homogeneous nebulae). In these cases, the amount of light entering the spectrometer is obviously proportional to the projected slit width δ. As pointed out earlier, for a given spectral resolution and a given collimated beam (or grating size), δ must be decreased if the telescope aperture D_T is increased. In principle, the resulting light loss at the slit can be compensated for by the D_T^2 dependence of the light collecting power of a telescope. However, this requires that the focal ratios of the telescope and the collimator be reduced accordingly. At large telescopes, these values are often at their technical limits and cannot be decreased further. Therefore, for high-resolution spectroscopy of extended objects with grating spectrometers, small telescopes are often more efficient.

4.4.2 Effects of Slit Illumination

For the discussion in the preceding paragraphs, a (more or less) uniform illumination of the spectrograph slit was assumed. In practice, this is almost never the case. Stars and other point sources produce a seeing pattern or (in the case of diffraction-limited observations) an Airy pattern at the position of the slit. For extended objects, the illumination depends on the light distribution of the target. Only extended nebulae, the sky background, and calibration sources may result in an essentially homogeneous slit illumination. Because seeing and diffraction patterns tend to have a central peak, for point sources the effective range of incidence angles Δi is normally smaller than the value $f_C^{-1}s$ given by Equation 4.8. Therefore, for point sources the effective spectral resolution may be somewhat higher than given by Equation 4.10. If the exact value of the spectral resolution is important for the interpretation of spectral data, it must be determined empirically by observing sources with sharp spectral lines and the same brightness distribution as that of the scientific targets.

If the spectrograph slit is illuminated uniformly, or if the illumination is symmetric with respect to the slit center, the finite slit width has no effect on the observed wavelength. However, if the illumination is not symmetric (e.g., if a stellar image is not aligned with the slit center), the spectrum is shifted on the detector, which results in a wavelength error. From Figure 4.7 it is clear that the amount of the wavelength shift can be of the same order as the wavelength uncertainty $\Delta\lambda$, as defined in Equation 4.6. This "slit effect" is absent or much weaker in fiber-coupled spectrometers (see Section 4.5). Therefore, for high-precision wavelength or radial velocity measurements, fiber-coupled spectrographs are better suited than slit instruments. A homogeneous slit illumination can also be achieved by means of *image scramblers* (again, see Section 4.5). However, fiber links, or a combination of fibers and image scramblers, tend to give better results.

4.4.3 Pre-Slit Optics

Image Slicers

Because of the relation between slit width and spectral resolution explained in the previous sections, small slit widths are desirable for observations of stars and other point sources. Therefore, significant efforts have been invested into methods that make it possible to reduce the slit widths to values smaller than the seeing pattern, without losing light at the slit. Apart from the AO systems mentioned earlier, *image slicers* are the most common devices used for this purpose. These devices redistribute the light of a seeing pattern, or of the exit end of a fiber, in such a way that it can pass through a narrower slit. During the past decade, various different types of slicer optics have been suggested and realized. An example of a modern image slicer is presented schematically

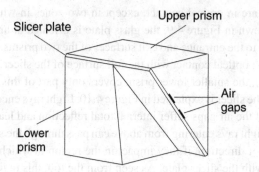

Figure 4.9. Perspective outside view of the two-beam image slicer described in the text.

in Figures 4.9 and 4.10. This particular slicer has been developed for a fiber-coupled spectrometer, in which it reduces the size of the image of the fiber ends in dispersion direction by a factor of 1/2, thereby doubling the spectral resolution. It slices the images of two fiber ends in exactly the same way. One fiber is always used to accept the light of the spectroscopic target. The second fiber can be used either to obtain simultaneously a spectrum of the sky background or of a comparison light source for wavelength calibration (see Figure 4.16). The principle of this slicer can also be applied to directly coupled spectrographs.

Technically, the slicer shown in Figures 4.9 and 4.10 consists of a plane-parallel thin glass plate between two prisms of the same material. The plate

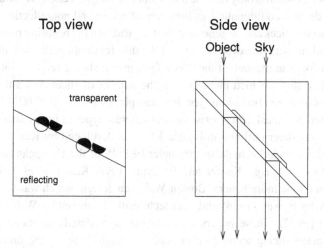

Figure 4.10. The optical principle of the two-beam image slicer.

and the prisms are in optical contact, except in two zones in which there are air gaps. As shown in Figure 4.9, the glass plate is oriented at an angle of 45 degrees relative to the entrance and exit surfaces of the two prisms. Whereas the upper prism is in optical contact with the full surface of the slicer plate (except for the air gaps), the smaller lower prism covers only part of this surface. The functioning of the slicer is explained in Figure 4.10. Light rays encountering the narrow zones of the air gaps suffer internal total reflection and leave the slicer. Otherwise, all light rays entering from above can pass through the slicer without loss or change of direction, if they impact in the region in which both prisms are in contact with the slicer plate. As seen from the top, this region is labeled "transparent" in Figure 4.10. Rays that impact the slicer outside this region (i.e., in the area labeled "reflecting" in Figure 4.10) suffer total reflection at the lower (glass–air) surface of the slicer plate, and are reflected by 90 degrees.

As shown in the left (top-view) part of the figure, the slicer is operated by imaging the fiber ends (or the seeing disk of an object) exactly on the edge of the reflecting zone. Therefore, one half of each image (indicated in black) passes through the device. The other half is reflected by 90 degrees toward the right. As indicated in the side-view part of Figure 4.10, another total internal reflection at an air gap at the upper surface of the slicer plate brings the light of the reflected half back into the original direction. Because the light is now in the "transparent" zone, it now also passes through the slicer, but it is displaced by about one diameter of the original disk. The same double reflection occurs for the sky fiber end. As shown by the figure, in this way the slicer converts each circular focal pattern into two half-moon-shaped images. The new pattern is about twice as long, but its width is only half the diameter of the original circle. By adding (in different depths) additional edges between reflecting and nonreflecting zones, more than two slices can be generated, and the width of the resulting pattern can be reduced further. However, because of the different light paths, not all slices can be in focus at the same time. Therefore, image slicing tends to introduce optical aberrations, which increase with the number of slices. For a thorough discussion of these aberrations, see, for example, Dekker et al. (2003).

As noted previously, there exist various different types of image slicers. The double-beam slicer outlined in Figure 4.10 was developed by Kaufer (1998) for the fiber-coupled echelle spectrometer FEROS of the European Southern Observatory (see, e.g., Kaufer and Pasquini, 1998; Kaufer et al., 1999). It was based on the more general Bowen-Walraven design, which was suggested originally by Bowen (1938) and later technically improved by Walraven und Walraven (1972). These papers also provide more details on these devices and on image slicing techniques in general. Because the slicing takes place at the focal plane, the slicer components must be small but of high optical quality. In the case of FEROS, the thickness of the slicer plate is only 0.2 mm.

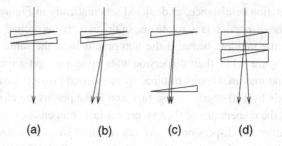

(a)   (b)   (c)   (d)

Figure 4.11. Different concepts of atmospheric dispersion compensators.

Manufacturing good image slicers, therefore, is among the most demanding tasks in constructing optical spectrometers.

Atmospheric Dispersion Compensators

Apart from blurring images (the "seeing effect"), the Earth's atmosphere also causes a chromatic dispersion of the light from astronomical objects. This dispersion occurs for all targets, except for objects in the exact zenith position. It is caused by the decrease of the density and the change of the composition of air (and the resulting change of the refractive index) as a function of the altitude in the atmosphere. Therefore, all inclined light paths in the atmosphere become curved, and the curvature depends on the wavelength. If bright stars are observed with a large telescope at a high zenith distance, the atmospheric dispersion can be seen directly by the eye. At a zenith distance of 50 degrees, the position of the blue and the red images of a star differ by about two arcsec, which (under good atmospheric conditions) is significantly larger than the seeing pattern and much larger than the diffraction pattern of a big telescope. To avoid an image degradation due to the dispersion effect, many modern telescopes are equipped with *atmospheric dispersion compensators* (ADCs), which (at least partially) eliminate this effect. If an ADC is not part of the telescope optics, it is sometimes included in spectrometers, in which it is installed in front of the spectrograph slit.

To eliminate the atmospheric effect, an ADC produces a dispersion of the convergent telescope beam, which just compensates for the atmospheric dispersion. Figure 4.11 shows examples of optical concepts that are being used for this purpose. The most simple ADC is a thin prism with a dispersion in the same plane and with the same amount as, but of opposite sign than, the atmospheric dispersion. However, a simple prism produces a fixed amount of dispersion. Thus, it can compensate for the atmospheric dispersion only at one given zenith distance. In a practical ADC, the dispersion must be adjustable. This can be achieved by combining two identical prisms with an adjustable relative orientation or distance, as depicted schematically in Figure 4.11. One way to vary the dispersion is to rotate one of the two prisms around an axis that is parallel to the incident beam. If the two prisms have the same orientation (case (b) in Figure 4.11), their dispersion adds up and we get a maximal total dispersion (and maximal compensation). If the second prism is rotated around the optical axis by 180 degrees (case (a)), and if the prisms are close together or in contact, the dispersions of the two prisms just compensate for each other, and the resulting net dispersion is zero (as required for zenith observations). Orientations between these two extremes give intermediate effective dispersion (and compensation) values.

A disadvantage of the type of ADC outlined by Figure 4.11 (a) and (b) is a tilt of the resulting corrected beam. This tilt, which depends on the prism rotation angle, shifts the images, and it tends to cause aberrations.

As indicated by case (c) of Figure 4.11, an adjustable dispersion can also be achieved by changing the distance of two prisms. In this case, the direction of the beam remains unchanged. However, the beam suffers a slight displacement, which again can cause aberrations.

An ADC that avoids a tilt *and* a beam displacement is shown as case (d) of Figure 4.11. This compensator consists of two identical Amici prisms. Each of these Amici prisms consists of two (or more) normal prisms of glasses with different refractive indices. By selecting glasses with suitable properties, for one (central) wavelength the tilt introduced by the first component of the Amici prism is canceled by the second prism, but the combination still produces a chromatic dispersion. The effective dispersion of the Amici prism pair is again adjusted by rotating one prism relative to the second one around the optical axis.

Although the Amici prism pair causes the lowest optical aberrations, the need to find suitable glasses tends to restrict their operational wavelength range. An ADC according to case (c) normally can cover a larger wavelength range if a prism material such as fused quartz is used.

The use of an ADC is not always of advantage. Independent of the design of an ADC, atmospheric dispersion is never completely eliminated, as glass never matches the wavelength dependence of the atmospheric refractive index exactly. Moreover, even with antireflection coatings, ADCs always result in some light losses.

4.4.4 The Littrow Configuration

During the history of spectroscopy, many different variations and improvements of the basic slit spectrometer of Figure 4.1 have been developed. One of the most important concepts is a configuration that was suggested originally in the

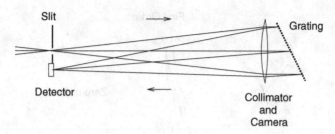

Figure 4.12. Schematic light path in a Littrow spectrometer.

nineteenth century by the Austrian astronomer Otto von Littrow (1843–1864) (see Littrow, 1863). Perhaps the best-known modern use of Littrow's concept is its application in external cavity diode lasers (ECDLs, "laser diodes"). Littrow's design is also still of significant importance in astronomy. As outlined in Figure 4.12, in the Littrow configuration the same optical element is used as collimator and as camera optics. The disperser usually is a reflection grating, which is illuminated very close to its blaze angle. However, it is also possible to realize a Littrow configuration with a grism or a prism followed by a plane mirror, or by a prism with a reflecting back face. Because the dispersed beam has (apart from the sign) nearly the same direction as the input beam, the slit and the detector must be placed closely together, or an additional reflection must be added.

Apart from the fact that no extra camera is needed, Littrow's design has the important advantage that optical aberrations, which are caused by the collimator, are largely compensated for during the return pass through the same optical element. Thus, high-quality spectra can be obtained with relatively simple optics. In astronomy, this is particularly valuable, as it avoids the light losses of more complex optical systems. Because blazed reflection gratings are most efficient when operated at their blaze angle, the Littrow configuration also results in the highest possible efficiency attainable with a given grating. A disadvantage of Littrow's arrangement is that the camera focal ratio is determined by the input beam. Therefore, the Littrow design is often combined with the white pupil concept and an additional detector-optimized camera.

4.4.5 Rowland Circles

Another particularly simple grating spectrograph is shown in Figure 4.13. This configuration was suggested by Henry Rowland (see Rowland, 1882). It has only one optical element, which is a reflection grating with a spherical concave surface and a curvature radius R_G. A simple calculation (see, e.g.,

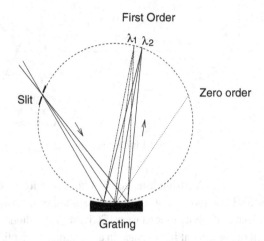

Figure 4.13. Schematic layout of a Rowland circle.

Palmer and Loewen, 2005) shows that the grating can produce sharp spectra if the spectrograph slit and the detector are placed on a circle with radius $R_G/2$, and if this circle touches the concave grating at its center (as indicated in Figure 4.13). In this case, the concave grating combines the functions of collimator, disperser, and camera. Spectrometers based on Rowland's design (usually called *Rowland circles*) are simple and inexpensive, but they cannot be used at high diffraction orders. Moreover, their resolution is limited by astigmatism and other optical aberrations. By using concave gratings with more complex curved surfaces and special groove patterns, the astigmatism can be reduced, but not fully prevented. Such aberration-corrected concave reflection gratings can be produced using the holographic technique mentioned in Chapter 3.

A disadvantage of the Rowland design is that the spectra are formed on a curved focal surface. During the photographic era, this problem was solved by using bent photographic plates or films mounted with the required curvature. With modern solid-state detectors, matching the curved focal plane is more difficult and usually requires expensive custom-made devices.

Rowland used his design successfully for precisely measuring the lines of the solar spectrum. As a result, large Rowland circles were installed at many solar observatories. However, for the reasons noted previously, they are rarely used for stellar spectroscopy. An exception is spectroscopy at extreme ultraviolet (EUV) wavelengths, at which the strong absorption and low reflectivity of all optical materials rule out the use of conventional collimators and cameras. As discussed in Chapter 7, at these wavelengths ($\lambda < 110$ nm) variations of

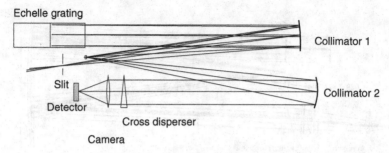

Figure 4.14. Schematic layout of an echelle spectrometer using the Littrow concept for the main dispersion. The divergent beam from the slit is collimated by means of a concave mirror. The echelle grating is inclined to the plane of the figure, and is shown in projection. The dispersed beam is refocused and reflected to a second collimator, which produces a white pupil at the cross disperser.

Rowland's design are still employed regularly. Another example of the use of Rowland's principle in present-day instrumentation is the waveguide array spectrometers discussed in Chapter 10.

4.4.6 Echelle Spectrometers

Echelle Optics

Practically all present-day high-dispersion spectrometers for optical wavelengths use echelle gratings and cross dispersers. The cross dispersers are either gratings, grisms, or prisms. Cross disperser gratings generally are used in the first order. If the spectrometer is designed to cover a large wavelength range, prism cross dispersers are preferred, because prisms are free of order overlaps. Echelle gratings typically have blaze angles $\geq$ 63.5 degrees and (depending on the operating wavelength) grating periods g between 3 µm and 40 µm. Most echelle spectrometers operate at diffraction orders $m > 40$, with typical values near $m \approx 100$. According to the grating equation (Equation 3.21), for a typical grating period of 32 µm and a blaze angle 63.5 degrees, $m = 100$ corresponds to a central wavelength of 573 nm.

Some contemporary echelle spectrometers use large Schmidt cameras to image the wide cross-dispersed beam to a detector. Notable examples are the HIRES spectrograph at the Keck Observatory (see Vogt et al., 1994) and the SOPHIE spectrometer at the 1.9-m telescope of the Observatoire de Haute-Provence in France (Perruchot et al., 2008). However, most current high-resolution instruments use a combination of the Littrow concept and the white pupil design. An example of this type of optical layout is given schematically in Figure 4.14. In this layout, the divergent beam from the spectrograph slit is

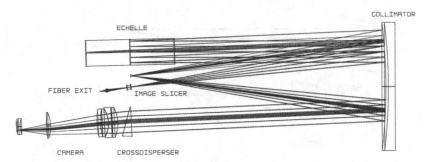

Figure 4.15. Optical layout developed for the short-wavelength (0.5–1.0 μm) spectrometer of the CARMENES radial velocity project of the Calar Alto Observatory (CAHA) in Spain. (Reproduced with permission of Wenli Xu Optical System Engineering.)

made parallel using an off-axis parabolic mirror. Following the Littrow concept, the same mirror functions as a camera to produce an intermediate spectrum. The resulting main dispersion and the intermediate spectrum are oriented perpendicular to the image plane of Figure 4.14. A small flat mirror (elongated in the direction of the main dispersion, i.e., perpendicular to the image plane of Figure 4.14) is used to reflect the beam to a second collimator mirror, which has the same optical parameters as the first collimator. To direct the beam to the flat mirror, the grating is slightly tilted. Because it is located close to the focal plane of the collimator, the flat mirror can be small. Because of its small surface, this mirror has only a very small cross section for the undispersed stray light, which is produced by the echelle grating and at the collimator. Thus, the reflection at this mirror removes most of the stray light from the beam. The second collimator produces a white pupil at the cross disperser. A camera behind the cross disperser then records the spectrum. For Figure 4.14, an "in-line" cross disperser has been assumed. If the cross disperser changes the beam direction, the camera must be tilted accordingly.

A more realistic layout of a modern echelle spectrometer, which follows the scheme outlined in Figure 4.14, is reproduced in Figure 4.15. This figure is based on a design study for the short-wavelength spectrograph of the CARMENES instrument, which is being developed for the Calar Alto Observatory in Spain (see Quirrenbach et al., 2010). Figure 4.15 is scaled and shows the actual optical components. A feature in which the CARMENES design differs from the schematic layout of Figure 4.14 is that in Figure 4.15, the two collimators have been merged into one single large concave mirror. This provides additional mechanical stability and simplifies the alignment of the instrument. The camera consists of a fully *dioptric* system (containing lenses only, with no reflective optical components). As can be seen from the figure, a relatively complex arrangement of lenses is required to get a good image quality (and good optical resolution) over the whole two-dimensional echelle pattern at all wavelengths.

Figure 4.15 assumes a 12.7 cm × 50.8 cm echelle grating with a 32-μm grating period and and a blaze angle of 76 degrees (R4). The cross disperser is a grism.

Current echelle spectrometers use either lens systems (as in Figure 4.15) or mirrors, or a combination. Reflective optics has the advantage of being fully achromatic and usable over a wider wavelength range. However, wide-field mirror optics usually results in shadowing effects and light losses due to imperfect reflection. Lenses have the disadvantages of chromatic aberrations and reflection losses at the air–lens surfaces. Moreover, their bandwidth is limited by the transparency of the glasses. On the other hand, using antireflection coatings, the reflection losses of dioptric systems can be kept low, and the many design parameters resulting from the wide choice of commercially available optical glasses make it easier to optimize dioptric systems. Therefore, for the wavelength bands, for which suitable refractive optical materials can be obtained, dioptric systems are usually preferred.

Like most modern high-resolution spectrometers, CARMENES will be a stationary instrument, coupled to the telescope using an optical fiber (see Section 4.5). In most cases, the optical components of such stationary spectrometers are mounted on stable optical benches made of a high-tensile metal or of granite. To avoid wavelength instabilities due to temperature and air pressure variations, modern high-resolution spectrometers often are enclosed in pressure-controlled vessels or (as in the case of the CARMENES instrument) in vacuum tanks. To avoid influences of the environment, such instruments are thermally and mechanically insulated and well shielded from all outside effects.

As noted earlier, in fiber-coupled instruments spatial information on the target is lost, whereas slit spectrometers provide at least spatial information along the projected slit. However, echelle instruments are not well suited for long-slit spectroscopy anyway, as the echelle format severely constrains the feasible slit heights. High-resolution spectrometers, which to some extent combine the high stability and wavelength accuracy of stationary instruments with limited spatial information along the slit, are Nasmyth-focus echelle instruments. Notable examples are the HIRES spectrograph at the Keck observatory (Vogt et al., 1994) and the UVES spectrograph at the ESO VLT (Dekker et al., 2000). At an alt-az mounted telescope, the Nasmyth focus position moves with the azimuthal motion of the telescope, but a spectrograph at this focus always feels the same direction of the Earth's gravitational acceleration. Thus, if the spectrograph is properly mounted, mechanical bending effects due to the telescope's movement tend to be small. On the other hand, at this focus the field of view rotates during observations. Therefore, if spatial information is important, a field "de-rotator" must be employed. Such devices are available at all large alt-az telescopes, but they always result in some light losses. Therefore, they are not used if only point sources are observed.

Gratings, collimators, spectrograph cameras, and detectors usually work most efficiently in a narrow spectral range. Instruments designed for a wide wavelength coverage always require compromises concerning the optical performance and efficiency. To retain a high efficiency at all operating wavelengths, modern high-resolution spectrometers are often split into two or more "arms," which are optimized for different wavelength bands. The different configurations can be used either alternatively (e.g., at HIRES) or, using dichroic beam splitters, simultaneously (e.g., UVES). If dichroic beam splitters are used, some wavelengths near the transition wavelength may not be observable. In this case, a full spectral coverage requires the presence of two or more interchangeable dichroic beam splitters.

A useful list of high-resolution spectrometers that are operated at present at large telescopes, or that are under construction, has been published on the CARMENES Web page (http://carmenes.caha.es). This list includes links to the Internet pages of the individual instruments, on which much additional information on these instruments and on echelle spectroscopy in general can be found.

Properties of the Echelle Format

Section 3.3.3 pointed out that the illumination pattern that an echelle spectrometer produces on the detector resembles the rungs of a ladder, and that the term *echelle* goes back to the French word for *ladder*. Figure 4.16 shows an observed example of this pattern. This spectrum was obtained with ESO's FEROS spectrometer, which uses a prism as cross disperser. Because the prism dispersion is nonlinear, the orders are curved.

A schematic example of the echelle format, showing some of the properties of the echelle pattern, is reproduced in Figure 4.17. The figure assumes a linear main and cross dispersion. This is normally a good approximation if gratings are used. For a first-order cross-disperser grating, we get from the grating equation

$$\lambda = g(\sin i_C + \sin j_C), \tag{4.12}$$

where the i_C and j_C are, respectively, the angles of the incident and the diffracted beams (relative to the grating normal) at the cross disperser. For small angles, this results in linear relation between λ and j_C.

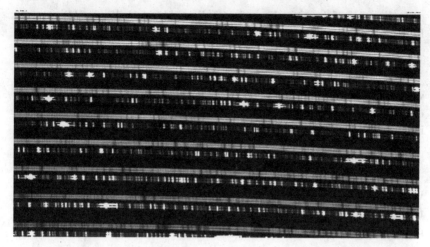

Figure 4.16. The central orders of an echelle spectrum. For each order, the target spectrum (showing stellar absorption lines) and a wavelength calibration spectrum (showing Th and Ar emission lines) are visible. Each spectrum shows two slices produced by the image slicer. From Kaufer et al. (1999).

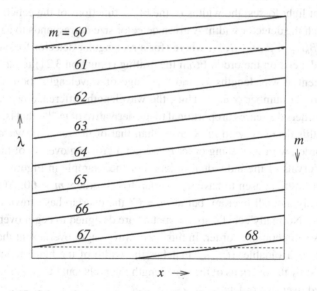

Figure 4.17. Schematic geometry of the orders of an echelle spectrogram. The horizontal broken lines indicate constant wavelengths. The position of a given wavelength on the detector is assumed to depend linearly on the wavelength.

In direction of the main dispersion (indicated by the subscript M), we get from the grating equation

$$m\lambda g^{-1} = \sin i_M + \sin j_M = \sin i_M + \sin(i_M + \beta), \qquad (4.13)$$

where β is the angular difference between the blaze direction and the direction of diffraction for a given wavelength and order. For small values of β, we have in linear approximation

$$m\lambda g^{-1} = \sin i_M + \sin i_M \cos\beta + \cos i_M \sin\beta \approx 2\sin i_M + \beta \cos i_M = a + b\beta, \qquad (4.14)$$

where a and b are constants. Thus, if the coordinate x describes the position on the detector in the direction of the main dispersion and relative to the detector center, for small angles β we get (again in linear approximation)

$$x \approx \beta f_{Camera} \propto (\lambda - \lambda_c), \qquad (4.15)$$

where λ_c is the central wavelength of the corresponding order.

The range of wavelengths that can be imaged at each order obviously depends on the width of the detector and the width of the blaze function. To prevent light losses, the widths of the blaze functions of the echelle orders must match the detector width. A given detector size corresponds to a range of angles $-\beta_{Max} < \beta < +\beta_{Max}$, where (for a rectangular detector format) β_{Max} does not depend on the order. From the grating (Equation 3.21) it can be seen that the central wavelengths λ_c and the range of wavelengths both decrease with the order number $\propto m^{-1}$. Thus, the wavelength difference between two adjacent orders decreases proportional to the derivative of m^{-1} – that is, $\propto m^{-2}$. Because this decrease with m is faster than that of the range of wavelengths of each order, with increasing order m we get a growing overlap of the wavelength intervals of the individual orders. For the example in Figure 4.17, the parameters were chosen to have no overlap for all orders $m < 60$. At $m = 60$ there is only a small overlap, but at $m = 67$ the overlap has grown to about 25 percent. Normally echelle spectrometers are designed to begin overlapping at the lowest operational order. In this case, wavelength overlaps at the higher orders are unavoidable. If β_{Max} matches the width of the blaze function, no light is lost by the overlaps of the wavelength intervals, but the power is simply distributed over two orders.

Technically, the onset of the wavelength overlap depends on the design parameters (g, β_{Max}, the range of orders, and the cross disperser). If a prism cross disperser is employed, the wavelength dependence of the dispersion of optical glasses can be used to optimize the wavelength distance between echelle orders. In spectrometers in which wavelength gaps occur at low echelle orders, a recording of the full spectrum requires sequential observations with two or more settings of the echelle grating.

The estimates described previously provide qualitative constraints for the construction and the use of echelle spectrometers. As noted, these estimates are based on linear approximations. Moreover, the relation between wavelengths and position on the detector is also affected by aberrations and distortions introduced by the optical components. Thus, exact relations must be calculated using optical design software, which models all optical components. Alternatively, the linear dispersion can be determined empirically using well-calibrated standard spectra.

4.4.7 Echelon Spectrometers

As noted in Section 3.3.7, echelon gratings are used in astronomy for some special applications. Because of the small free spectral range of the echelon gratings, these devices always must be used with cross dispersers. The spectral dispersion required for separating the orders of an echelon often can be achieved only by using echelle gratings as cross dispersers. Therefore, such spectrometers typically combine an echelon grating and an echelle grating (used as cross disperser). An example of this special type of high-resolution instruments is the mid-IR spectrometer EXES on the Stratospheric Observatory for Infrared Astronomy (SOFIA; see Section 4.8.4). A detailed description of EXES and references to other devices of this type have been given by Richter et al. (2003).

4.5 Fiber-Coupled Instruments

The principle of fiber-coupled spectrometers was illustrated in Figure 4.5, and some examples have already been mentioned in the preceeding sections. Using flexible fibers, these instruments can be installed stationary at some distance from the telescope, whereas only a small fiber-feed unit has to be mounted at the telescope itself. Fiber-coupled spectrometers normally have the same optical layout as slit spectrometers, except for the slit, which is replaced by the output end of the fiber. As pointed out in Section 4.4.3, instead of using the fiber output directly, to increase the spectral resolution, the fiber end can be imaged onto an image slicer.

Optical fibers are manufactured mainly for telecommunication or illumination purposes. Most common are *single-mode* fibers, with core diameters of the order of the operating wavelength. These fibers are the preferred choice for telecommunication, but (except for very special applications; see Section 10.3.3) they are not suited for feeding astronomical spectrographs. On the other

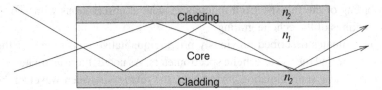

Figure 4.18. Cross section and light paths of a multimode optical fiber.

hand, *multimode* fibers, with core diameters of the order 100 μm, are being used regularly in astronomical spectroscopy at visual wavelengths. The functioning of a multimode fiber, as outlined in Figure 4.18, can be understood from geometric optics. In principle, an optical fiber is a very long (relative to its diameter) glass cylinder. Normally the cross section perpendicular to the fiber axis is circular, but fibers with rectangular or elliptical cross sections are also available, and are sometimes of advantage in astronomy. As indicated in Figure 4.18, a fiber always has a *core* of glass (or another transparent material) with a refractive index n_1, and a *cladding* with an index $n_2 < n_1$. The index may change discontinuously (in a step-index fiber) or gradually (in a gradient fiber). If light enters the core at a sufficiently small angle to the fiber axis, it will experience total internal reflection at the interface between the core and the cladding. This reflection will be repeated each time the light encounters the interface zone. The light losses for internal total reflection are very small. However, some flux attenuation occurs as a result of Rayleigh scattering at the glass molecules and by absorption in the glass or by impurities. Because the attenuation by Rayleigh scattering is proportional to λ^{-4}, this effect is most important at short wavelengths, whereas molecular absorption becomes critical in the IR. Because of the scattering losses, optical fibers are rarely used at UV wavelengths. In the IR, selected fibers are being employed for wavelengths up to about 2 μm (e.g., in the NIR spectrometer of the CARMENES project). New fibers that have become available recently may possibly extend the wavelength range to about 4 μm, if thermal emission effects can be overcome.

A given light beam generally leaves the fiber with an angle to the exit surface that differs from the angle it had relative to the entrance surface. However (assuming an ideal cylindrical geometry and plane entrance and exit surfaces, which are perpendicular to the fiber axis), the angle to the fiber axis does not change and will be the same at the fiber input and at the fiber exit (see Figure 4.18). Thus, if the light rays entering an ideal fiber have a range of angles $\alpha < \alpha_{max}$ to the optical axis, the same range of angles is present at the fiber exit.

At a telescope focus, this range of angles is given by the relation

$$\tan \alpha_{max} = \frac{D_T}{2 f_T}, \qquad (4.16)$$

where f_T is the focal length and D_T the aperture diameter (or, more exactly, the diameter of the entrance pupil) of the telescope. This relation contains the inverse of the *focal ratio* or *f -number* f_T/D_T of the telescope. Thus, an ideal fiber conserves the focal ratio of the input beam.

Total internal reflection is known to occur only when the angle α to the fiber axis meets the condition $\cos \alpha > n_2/n_1$. For larger angles α, the light is not reflected, but it penetrates into the cladding, where it is lost. Thus, a fiber transmits only light with a certain range of angles α. This range is called the *acceptance cone*. Naturally, the acceptance cone depends on the fiber material, but typical limits are $\alpha \leq 20$ degrees.

Figure 4.18 assumed a completely linear geometry. Because thin glass fibers are flexible, they are normally not exactly straight. From Figure 4.18 it is clear that any bending of the fiber may change the internal reflection angles, which may result in different exit angles and a decrease of the acceptance cone. However, as long as the curvature radius of the fiber remains much larger than the fiber diameter, these effects remain unimportant. Often a fiber will break before the optical effects of the bending become significant. Nevertheless, for mechanical as well as for optical reasons, strong curvatures of fiber lines should be avoided during the design and the operation of fiber-coupled spectrometers.

Variations of the core diameter or variations of the refractive index have the same effect as bending a fiber. Although modern manufacturing techniques produce fibers with surprisingly uniform diameters, there are deviations from the ideal geometry, which result in an increase of the range of angles α at the fiber output relative to the input. Thus, although (as explained earlier) in an ideal fiber the focal ratio remains unchanged, in real fibers the exit focal ratio tends to be lower than that at the fiber input. This effect (known as *focal ratio degradation*) is most critical for large values of f_T/D_T, where at the input all rays have small angles α. Therefore, to avoid a significant change of the focal ratio, the convergent beam from a telescope is often converted to a smaller focal ratio (or wider range of angles) before it is fed into a fiber. Behind the fiber the original focal ratio is restored, or f_T/D_T converted to a value that is convenient for the spectrometer optics. The focal ratio conversion can be carried out by reimaging the focal plane, or by cementing microlenses onto the fiber ends.

Because at the fiber output the directions of light rays are (apart from the angle to the fiber axis) unrelated to the input directions, spatial information on the target is lost. Although this may be a disadvantage for some applications, the scrambling of the light rays is quite helpful for high-accuracy wavelength measurements and radial velocity observations. As pointed out in Section 4.4.2, in the case of slit spectroscopy, wavelength measurements are affected by the position and light distribution of the target image in the slit. In fiber-coupled instruments, the light distribution at the fiber input has only a minor effect on the light distribution on the detector and, therefore, almost no effect on wavelength accuracy of the measurement. Only in the case of very precise radial velocity measurements additional image scramblers or image homogenizers may have to be included in the fiber link. These image scrambles can be, for example, fibers with variable diameters. Detailed descriptions and discussions of image scramblers have been given by Avila and Singh (2008) and Barnes and MacQueen (2010).

4.6 Multiobject Spectrometers

For surveys and for other programs in which many objects of interest are present in the field of view of a telescope, valuable observing time can be saved by taking the spectra of several (or many) targets simultaneously. Methods that permit such observations are referred to as *multiobject spectroscopy* (MOS). One example – slitless spectroscopy – has been presented in Section 4.3. As described in that section, slitless spectra of multiple objects can be obtained using objective prisms in front of a telescope, or by using the basic layout of a slit spectrometer without a slit in the telescope's focal plane. Although the objective prism technique is restricted to small telescopes, the latter version can be applied to telescopes of any size. However, as explained in Section 4.3, in ground-based astronomy the sky background limits slitless spectroscopy to relatively bright objects. Moreover, overlapping spectra can lead to a loss of information. Obviously, these problems can be avoided by placing an individual slit at the image of each target in the focal plane of a telescope. Alternatively, multiple fiber feeds can be installed in the focal plane. Both these concepts are frequently applied in present-day astronomical spectroscopy. In the following paragraphs the techniques are described that are currently in use for multiobject spectroscopy.

4.6.1 Multislit Spectroscopy

A schematic optical layout of a multislit spectrometer is presented in Figure 4.19. For clarity, this example shows slits for three targets only. However, the concept can obviously be extended to a much larger number of objects.

As noted earlier, MOS slits must be arranged in the focal plane of the telescope. To prevent overlaps of the spectra, the individual slits normally are staggered in the direction perpendicular to the dispersion (i.e., in Figure 4.19, perpendicular to the image plane). A grating or prism is placed where the collimated beams from the individual slits form a common pupil. A camera then focuses the dispersed beams onto a detector, on which a spectrum is recorded

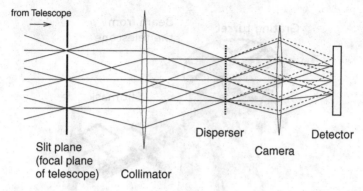

Figure 4.19. Schematic optical layout of a multislit spectrometer. This particular layout assumes an in-line disperser (such as an in-line grism). The spectral dispersion is assumed to be in the figure plane, and the slits are assumed to extend normal to this plane. Only one open slit is shown to indicate that only this slit is in the figure plane, whereas the other slits are above or below to prevent overlaps of the spectra.

for each slit. If the disperser is removed from the beam (or replaced by a mirror in the case of a reflection grating), and the slits are completely opened, the instrument can be used as a camera to image the sky. Compared with a direct imaging in the focal plane of the telescope, the configuration of Figure 4.19 has the advantage that the image scale can be adjusted to the detector size and to the detector resolution. Moreover, the collimator and camera optics can be used to improve the image quality over the field of view. Because the camera focal length normally is shorter than that of the telescope, the arrangement of Figure 4.19 without disperser and slits is called a *focal reducer*.

In Figure 4.19, the disperser is assumed to be a transmission grating or a grism. However, multiobject spectrometers can also be based on reflection gratings and mirror optics. An example of the optical layout of an MOS spectrometer that (apart from Schmidt corrector plates) contains reflective optics only is given in Figure 4.20. There also exist MOS instruments that combine reflecting and dioptric optics. A notable example is the Low-Resolution Imaging Spectrograph (LRIS) at the Keck observatory (see Oke et al., 1995).

Although MOS slit assemblies based on micro-electromechanical systems (MEMS) are under development at various laboratories (see, e.g., Canonica et al., 2010; Li et al., 2010), most current multislit instruments use either individually movable slit jaws, as in the ESO FORS instruments (Appenzeller et al., 1998) and the Keck instrument MOSFIRE (McLean et al., 2010), or exchangeable *slit masks*, as in the case of LRIS at Keck, FOCAS at the Subaru telescope (Kashikawa et al., 2002), MODS, and ESO's FORS2. Movable slit

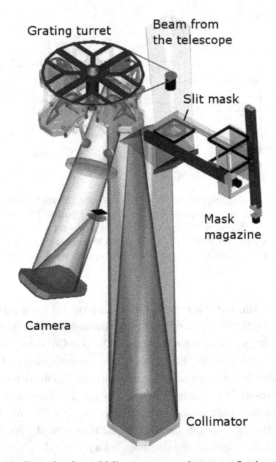

Figure 4.20. Example of a multislit spectrometer that uses reflective optics. The figure shows the layout of the red arm of the Multi-object Double Spectrograph (MODS) at the Large Binocular Telescope (see Osmer et al., 2000). The convergent telescope beam enters from above and reaches an exchangeable slit mask. Two folding mirrors (one of which is acting as a dichroic beam splitter) send the beams from the slits to a collimator mirror. The dispersion occurs at one of the exchangeable reflection gratings, which are mounted in a grating turret. The spectra are recorded by an off-axis Schmidt camera. Reproduced with the permission of the MODS team at Ohio State University.

jaws make it possible to adjust the slit widths to the seeing conditions at the time of the observations, and the resulting slit properties are more reproducible. Slit masks provide more flexibility concerning the slit arrangements and the possibility of using tilted slits (to avoid adjacent background objects in a target slit). However, the slit masks must be produced off-line and in advance of

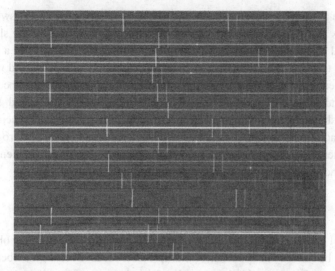

Figure 4.21. Section of a spectrum observed with the ESO FORS1 multislit spectrometer. The horizontal bright lines are the target spectra. Because all targets are hot stars, only a few and weak stellar absorption lines are visible in the (red) spectral range reproduced in the figure. However, most spectra show conspicuous emission lines of Hα. The bright vertical lines are [OI] night-sky (airglow) features. These lines show the height of the slits, which had a projected length of 18 arcsec. In some slits (by chance) several objects are present. Image courtesy ESO/FORS team.

the observations. Multislit masks can be manufactured mechanically or by means of computer-controlled laser cutting devices. For ambient temperature instruments, low thermal expansion materials (such as Invar) are used for the mask blanks. In temperature-controlled cooled spectrometers, aluminum or steel is adequate. After cutting, the masks are stored in a mask magazine, from which they are inserted into the focal plane of the telescope by a robot when needed during the observations (see, e.g., Figure 4.20).

Slit jaws that are manufactured with precision mechanical machines can be made to have very smooth edges. However, mask slits produced by laser cutting have been found to show edge irregularities, which can result in local slit width variation of up to a few percent. This can lead to an incorrect background subtraction and to errors in the extracted spectra of faint objects. A careful flat-field correction can eliminate this effect during the reduction procedure (see Section 6.4.3). However, in such cases even small systematic errors of the flat-fielding tend to produce residuals in the final spectra.

Figure 4.21 shows an example of a multiobject spectrogram. Because a spectral region near Hα in the spectra of hot stars is reproduced, few absorption lines are visible. However, several of the observed stellar spectra show conspicuous Hα emission lines. Because of the different positions of the slits in the focal plane, Hα occurs at different x coordinates, but always at a fixed distance to the [OI] airglow lines (which are present on all spectra, and which can be recognized by their characteristic pattern). For each slit, the spectrum above and below the target can be used to determine the sky background. However, as shown by the figure, in deep spectroscopic images there are practically always additional objects (stars and/or distant galaxies) in the spectrograph slits. Therefore, regions free of such background objects must be identified before the sky contribution can be determined and subtracted.

4.6.2 Multifiber Instruments

As noted earlier, instead of using multiple slits, spectroscopy of many objects can also be carried out by means of multiple fiber feeds. Because the fiber output ends can be arranged independently of the positions of the fiber inputs, the fiber feeds can be placed anywhere in the focal plane without risking spectral overlaps on the detector. Moreover, if needed, several different spectrographs can be connected to a single telescope focal plane. Thus, in a single exposure fiber-coupled MOS instruments can produce many more spectra than is technically feasible with multislit spectrometers. A notable example for the large multiplex gain, which can be achieved with fiber-coupled MOS, is the multiobject spectrographic system of the LAMOST telescope at the Xinglong observatory in northern China, which uses 4,000 fibers to feed 16 spectrographs (see Cui et al., 2008). To obtain the 4,000 spectra simultaneously, LAMOST uses, in total, about 130 km of optical fibers!

Fibers can also be used to do multiobject high-resolution echelle spectroscopy, which cannot be done efficiently with multislit instruments. An example for an instrument with a fiber-coupled echelle MOS capability is the FLAMES system at the European Southern Observatory Very Large Telescope (ESO VLT).

Several different methods can be used to place the input fiber ends at the target image positions in a focal plane of a telescope. Some multifiber instruments (such as LAMOST) use an individual actuator for each fiber. Other spectrometers (such as 2dF and 6dF at the Australian Siding Spring observatory) use robots to move the fiber feeds, which are attached magnetically to a steel base plate during the observations. To move the fiber feeds, the robot temporarily deactivates the magnetic field. The (so far) largest and most important multiobject spectroscopic survey program has been the Sloan Digital Sky Survey (SDSS; see York et al., 2000). For the SDSS, multifiber observations holes were drilled into "plug plates," using a computer-controlled milling machine.

Figure 4.22. Putting optical fibers ends into a plug plate for the SDSS multifiber spectrometers. Shown in the photo is Richard Kron, University of Chicago and Fermilab, and former director of the SDSS project. Image: Tom Nash, Fermilab.

Then the about 600 fibers of each spectroscopic exposure were inserted into the plug plates by hand (see Figure 4.22).

Although fiber-coupled MOS instruments can achieve larger multiplex gains and make it possible to reach higher spectral resolutions, they are less suited for the spectroscopy of faint objects with a surface brightness close to, or lower than, the sky background. Because the diameters of the fiber ends are not adjustable, relatively large fiber diameters are normally used to have a high throughput even at mediocre seeing. As a result, the sky contributions tend to be large. Moreover, whereas in multislit instruments the background can be determined close to the target in the same slit, in fiber-coupled instruments one must rely on "sky fibers" at some distance from the target, which cannot be easily checked for contamination by background objects. Finally, the unavoidable light losses in the fiber lines of large fiber-coupled MOS instruments reduce the sensitivity for faint objects. Therefore, for the spectroscopy of faint targets, multislit instruments are clearly preferable.

4.7 Integral Field Spectroscopy

As explained earlier, in conventional slit spectroscopy of extended objects, spatial information on the strength and the radial velocities of spectral features can be derived only along the direction of the projected slit. To obtain spatial information in the dispersion direction, several exposures with shifted projected slit positions must be made. To avoid such time-consuming sequential observations, many current spectrometers use image slicing techniques to obtain two-dimensional spatial information directly. These techniques are called *integral field spectroscopy* (IFS), and the optical devices that have been developed for this purpose are called *integral field units* (IFUs).

An image slicer was described in Section 4.4.3. In that section, image slicing was introduced as a method to split and to rearrange the seeing pattern at a telescope focus to match a narrower slit width and to achieve a higher spectral resolution. As described in Section 4.4.3, this is accomplished by slicing the seeing pattern into two or more parts. These slices are combined to form a narrower pattern at the spectrograph slit to form one single target spectrum.

In the case of IFS, the image of an astronomical object is sliced in a similar way, but the spectra of the individual slices are recorded and evaluated separately. Therefore, IFS uses different types of slicers. An example of an IFS slicer is shown schematically in Figure 4.23. An image of an actual slicer, which follows the principle of Figure 4.23, is reproduced in Figure 4.24. This particular slicer was developed for the near-infrared IFS device SPIFFI (see Eisenhauer et al., 2003) at the ESO VLT. It produces thirty-two slices. Because each slice results in a long-slit spectrum, the actual number of resulting resolved image elements is higher. If the period of an image slicer matches the spatial resolution, a slicer with a quadratic format producing n slices will result in n^2 spatial image elements. Thus, in principle, an instrument such as SPIFFI can resolve about 10^3 spatial elements.

Alternatives to reflective image slicers of the type described in Figure 4.23 are devices based on a combination of microlenses and optical fibers. The concept of this type of slicers is outlined in Figure 4.25. In principle, these IFUs correspond to a multifiber spectrometer with closely packed input fiber ends. To prevent light losses by light falling onto the fiber claddings or into gaps between the fibers, each fiber end is preceded by a microlens, which feeds the light into the core of the corresponding fiber. Using two-dimensional arrays of closely packed quadratic or hexagonal microlenses, gaps between the lenses can be avoided and a continuous field can be observed with only minor losses at the lens edges. Instead of two-dimensional microlens arrays, sometimes crossed arrays of long cylindrical lenses are used. As in the case of normal multifiber spectrometers, the fiber output ends can be distributed to several different spectrographs. Therefore, large numbers of spatial image elements, as well as high-resolution IFS, are possible with fiber-based IFUs.

At the time of this writing, the largest operational fiber IFU seems to be that of the VIMOS spectrometer at the ESO VLT (see Le Fèvre et al., 2002),

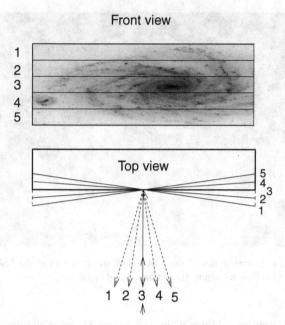

Figure 4.23. The principle of a simple (five-slice) IFS image slicer. It consists of a stack of five narrow mirrors with slightly different mirror normals. The beam from the telescope arrives in projection parallel to the normal of mirror 3 and forms an image on the front surface of the slicer, which divides the image into five slices. Because of the different mirror normals, the beams corresponding to the five slices (numbered 1–5) are reflected to different directions. A small tilt of the input beam relative to the image plane ensures that the output beams are above (or below) the input beam. Using additional reflections, the five beams are imaged to the slit plane of a spectrometer, in which the individual image slices are arranged to form one long slit.

which uses 80×80 spatial elements, producing 6,400 individual spectra. An example of spectra produced by this instrument is shown in Figure 4.26. Much larger fiber IFUs are under development. Among the most ambitious projects is VIRUS, a new IFU for the Hobby-Eberly Telescope (HET) in Texas, which (according to the present design) will include 145 individual spectrographs and produce 35,670 spectra simultaneously (see Hill and MacQueen, 2002). Such large slicers can perform tasks that normally are carried out by MOS systems.

Image slicing can also be carried out using fibers without microlens arrays, and with lens arrays without fibers. As noted previously, without microlenses, slicing with fibers results in light losses and in the loss of information about the image areas that correspond to the gaps between the fiber cores. The lost image information can be retrieved by means of additional, spatially shifted exposures.

Figure 4.24. Slicing optics of the NIR image slicer SPIFFI at the ESO VLT. Courtesy Max Planck Institute for Extraterrestrial Physics.

However, this requires additional observing time. Slicers with microlens arrays without fibers follow the principle of multislit spectrometers (see previous section). To avoid overlaps of the resulting spectra, the incident beams are tilted relative to the lens plane. (For details see, e.g., Bacon et al., 1995.) A well-known example of an IFS spectrometer of this type is the OSIRIS instrument of the Keck Observatory. A good description of this instrument and its optical layout is presented on the Keck Observatory home page (www.keckobservatory.org). Although image slicers that use microlenses without fibers have a high optical throughput, the constraints resulting from the arrangement of the spectra on the detector tend to result in an inefficient use of the detector area.

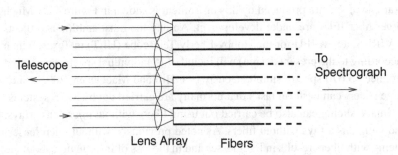

Figure 4.25. Schematic layout of an image slicer using microlenses and optical fibers.

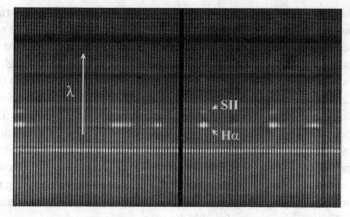

Figure 4.26. One hundred twenty (of 6,400) spectra produced by the ESO VIMOS instrument in IFS mode. The target is a planetary nebula showing Hα and [SII] emission lines. The horizontal features of constant brightness are due to [OI] air glow lines. The dark horizontal features are caused by A and B band atmospheric absorption. From Le Fèvre et al. (2002).

A fairly complete list of IFUs that are currently in use, or that are under development, can be found in the Internet at http://ifs.wikidot.com. This Web page also includes valuable links to further information.

4.8 Cold IR Spectrometers

The basic optical layouts for spectrometers described in the preceeding sections also apply to instruments for the infrared spectral range. However, at the normal temperatures of telescope enclosures, all objects radiate in the infrared. To avoid a thermal background from their optical and mechanical components, astronomical spectrometers for wavelengths $\lambda > 1.6$ μm usually are cooled to temperatures well below the ambient values. This requires modifications and specific technical features, which are discussed in this section.

4.8.1 General Principles

Any light entering a spectrometer at a ground-based instrument must pass through the Earth's atmosphere and the telescope. Both are warm, and both, therefore, unavoidably contribute some thermal background to the incident radiation. However, because (according to Kirchhoff's law) ideal reflectors do not emit thermal radiation, the contribution of the telescope can be minimized by making sure that, apart from sky light, only radiation from reflective surfaces (such as the primary and secondary mirrors) reaches the interior of the spectrometer. Contributions from the Earth's atmosphere cannot be avoided. At NIR wavelengths, the atmosphere contributes essentially emission lines ("airglow") and stray light (for details, see Glass, 1999). Apart from sharp absorption lines, in the NIR the air is rather transparent, and (owing to the decrease of stray light with the wavelength) the sky is very dark between the airglow lines. Thus, NIR spectra can be observed with high sensitivity if the spectral resolution is high enough to resolve the atmospheric lines and if the thermal background produced inside the spectrograph is kept low by cooling. Depending on the wavelength bands for which the instruments are designed, the operating temperatures of cooled spectrographs are between 77 K (achievable conveniently with boiling nitrogen) and values < 1 K.

Traditionally, IR spectrometers have been cooled by means of commercially available (or externally produced) liquefied gases (such as liquid helium or nitrogen). In most modern ground-based IR spectrometers the cooling is achieved with closed-cycle coolers, which work according to the same principle as household refrigerators, except that helium is used as refrigerant. For sub-Kelvin temperatures ^{3}He is required, or different cooling techniques (such as adiabatic demagnetization) must be employed. To reach low operating temperatures faster, the initial cool-down is frequently assisted using readily available liquid nitrogen.

Because of the high power requirements of closed-cycle coolers, and because of reliability issues, infrared space observatories sometimes still are cooled using an onboard supply of liquid helium. Because the helium evaporates slowly, in this case the amount of available liquid helium and the evaporation rate determine the lifetime of such missions. In addition to helium, some space experiments use solid neon as coolant, as solid neon has a much higher cooling power per unit mass.

4.8.2 Design Details

For thermal insulation and to prevent condensation of water and air, ground-based cooled spectrometers must be inclosed in vacuum vessels. Inside the cryostat, these spectrometers contain the same basic components as ambient-temperature spectrographs. However, because vacuum vessels are heavy and because the energy requirement for the cooling depends on the surface, the sizes of these vessels must be kept as small as possible. Therefore, in cooled instruments the light paths are arranged to minimize the space requirements. As a result, IR instruments often have complex and tightly folded optical layouts. The folding of beams is facilitated by the high IR reflectance of many mirror materials, which makes it possible to include many folding mirrors without significant light losses.

Because they are imaged onto the detector, cooling the slits of IR spectrometers is particularly important. As at other wavelengths, IR multislit spectrometers can be designed using individually movable slit jaws or slit masks. If movable slit jaws are used (as in the MOSFIRE NIR spectrometer of the Keck Observatory), the electromechanical components must be designed to work reliably under cryogenic conditions. If masks are used, it must be possible to insert the masks into the cold vacuum and to remove them again after the observations. Examples of NIR multislit spectrometers using exchangeable cold masks are MOIRCS at the Subaru telescope (Ichikawa et al., 2006) and LUCIFER at the Large Binocular Telescope (LBT) (Seifert et al., 2003, 2010). Figure 4.27 shows, as an example, the optomechanical layout of the LUCIFER spectrometer. The light from the bent Gregory focus of the LBT reaches the instrument horizontally (in Figure 4.27, from the left) at (1). It enters the vacuum vessel through the tilted dichroic window (2). At this window, the visual light is reflected to a wavefront sensing and guiding device, while the IR light forms an image on a slit mask just behind the entrance window. The slit masks are taken from the mask magazine (3) and placed into the focal plane by a robot. After passing the collimator optics (4) and folding mirrors (5), the beam is sent to a reflection grating in the grating turret (6). The resulting dispersed beam is reflected into one of the three cameras in the camera turret (7), and a spectrum is recorded by the detector (9). In front of the detector, a pair of filter wheels (8) can be used to select wavelength bands. By selecting a plane mirror instead of a grating in the grating turret, the instrument can be used for imaging and filter photometry. At (10) is an air lock, which is used to exchange the mask magazine. By means of this air lock, mask magazines can be inserted and removed without warming up the instrument. For this purpose, the masks are precooled in a separate vacuum tank, which is temporarily connected to the main vessel for the mask exchange (see Figure 4.28). The vacuum is kept by means of an integrated pump (11). The temperature is maintained by means of a closed-cycle cooler (not visible in the figure). To prevent heat input by conduction, the cold inner structure and its heat shield are connected to the outer vessel only by the plastic strips (12), which form a zigzag pattern to the left of the inner-structure base plate. These strips are designed to keep the structure accurately in position, while at the same time taking up the significant thermal stresses that occur during the cool-down of the spectrometer. Inside the outer wall (13) of the cryostat, a multilayer thermal shield prevents radiative heating of the spectrometer components by radiation from the outer wall.

Some cooled spectrometers are equipped with IFUs. In most cases, these are based on reflective image slicers (as in the case of the SPIFFI instrument

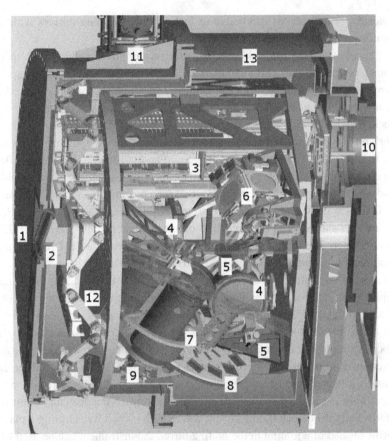

Figure 4.27. CAD image of the NIR multiobject spectrometer LUCIFER at the Large Binocular Telescope. The numbered components are explained in the text. Image courtesy LUCIFER team.

mentioned earlier) and are restricted to the wavelengths ≤ 2.5 μm. An exception is the instrument PIFS of the Mount Palomar Observatory, which can be used at wavelengths up to 5 μm. An innovative new NIR (1–2.4 μm) instrument is the spectrometer KMOS (Sharples et al., 2010) at the ESO VLT, which contains twenty-four individually deployable 14×14 pixel IFUS. Thus, KMOS combines MOS and IFS capabilities in a single NIR instrument.

4.8.3 Ground-Based MIR Instruments

According to Wien's displacement law, the wavelength λ_{max}, at which a blackbody radiation reaches the maximum of its intensity distribution, is given by

$$\lambda_{max} = \frac{2.8977685 \times 10^{-3}}{T} \text{m} \cdot \text{K}, \qquad (4.17)$$

Figure 4.28. Loading a magazine with precooled slit masks into the cold NIR spectrometer LUCIFER. The masks have been cooled down in the small auxiliary cryostat, which is visible in front of the larger spectrometer cryostat. For transferring the masks to the spectrometer, the two cryostats are temporarily connected via an air lock. For details of the mask exchange procedure, see http://www .mpegarching.mpg.de/ir/lucifer. Image courtesy LUCIFER team.

where T is the black-body temperature. For normal environmental temperatures, this maximum occurs in the mid-infrared near $\lambda \approx 10$ μm. Thus, cooling is particularly important for MIR and FIR observations. Moreover, above a wavelength of 2.5 μm the number density of atmospheric absorption lines increases significantly, making the atmosphere practically opaque (and strongly radiating) at wavelengths >40 μm. The atmosphere starts to become transparent again only at submillimeter wavelengths near 350 μm. IR observations above 40 μm, therefore, cannot be done from the ground. Even at wavelengths <40 μm, the lower thermal background makes MIR observations from space usually much more sensitive. On the other hand, with present technologies, space telescopes cannot reach the apertures of the largest existing ground-based instruments. Therefore, MIR observations with large ground-based telescopes can give higher spatial resolutions, which is of critical importance for some astrophysical applications. Thus, although most MIR spectroscopy is being carried out from space, some work is being done from the ground by making use of "wavelength windows" in which the atmospheric transparency and the background produced by the atmosphere are still acceptable. Table 4.1 gives approximate wavelength limits and approximate mean optical depths of some NIR and MIR windows (and their designations). The exact values depend on the

Table 4.1. *Approximate wavelength limits and mean optical depths of some atmospheric wavelength windows*

Spectral band	λ (μm)	Optical depth
J	1.1–1.4	0.03
H	1.5–1.8	0.02
K	2.0–2.5	0.03
L	3.0–4.1	0.15
M	4.6–5.0	0.30
N	7.5–14.5	0.08
Q	17.0–25.0	0.30

telescope site and on the meteorological conditions. Data on the atmospheric transparency at a given site can be found on the Internet pages of all major observatories. Particularly informative are the tables and graphs available at the Gemini Observatory Web page (www.gemini.edu) for the representative sites Mauna Kea (Hawaii) and Cerro Pachon (Chile).

The thermal emission of the atmosphere depends on the optical depth and the temperature. Up to the Q band, the atmospheric optical depth does not increase dramatically. However, for a typical effective temperature of the atmospheric radiating layers near 250 K, the steep slope of the Wien wing of the Planck function results in a very fast rise of the thermal background, which between the L band and the Q band increases by a factor $> 10^4$ (see, e.g., Léna et al., 1998).

A good example of a ground-based MIR spectrometer is the instrument MICHELLE of the Gemini observatory. MICHELLE covers the wavelengths 7 to 26 μm (N and Q bands). It can produce low-resolution spectra (R between 100 and 200), medium-resolution spectra ($R = 3,000$), and high-resolution echelle spectra ($R \leq 30,000$). (For a detailed description of this instrument, see Glasse et al., 1993.)

4.8.4 Space-Based and Airplane Observatories

As noted already, most MIR observations are carried out from space. Apart from the absence of atmospheric absorption and emission, space-based MIR observations have the fundamental advantage that, in addition to the spectrometer, the telescope optics can be cooled to a few degrees K.

Figure 4.29. The stratospheric infrared observatory SOFIA. Its 2.7-m telescope is located behind the half-opened shutter in the rear part of the aircraft. Image courtesy NASA/Jim Ross.

Important examples of recent space-based spectroscopic facilities with cold telescopes are the Spitzer Infrared Space Observatory (NASA) and the Herschel Space Observatory (ESA). Spitzer was launched in 2003. It carried the spectrometer IRS (Houck et al., 2004) for the wavelength range 5 to 38 μm and spectral resolutions R between 60 and 600. The higher resolution was achieved with an echelle spectrometer, operating at orders $11 \leq m \leq 20$. Using liquid helium, the temperature of the spectrometer was kept at 1.8 K, whereas the temperature of the 85-cm telescope was 5.5 K. The spectrometer was operated until 2009, when the helium supply became exhausted and the telescope temperature climbed to about 30 K. The Spitzer observatory is still being used for NIR photometry, but the spectrometer is no longer operational.

The Herschel Space Observatory was launched in 2009. Its 3.5-m main mirror is passively cooled to about 90 K. Among other instruments, it includes the grating spectrometer PACS for wavelengths between 57 μm and 210 μm (Poglitsch et al., 2008). The spectral resolution depends on the wavelength and varies between 1,000 and 4,000. PACS includes a small (5×5 element) IFU. Depending on the rate of helium evaporation, the Herschel observatory is expected to remain operational until 2013.

A valuable complement to ground-based and space-based IR spectrometers are instruments on stratospheric balloons and on airplanes. An example is the the Stratospheric Observatory for Infrared Astronomy (SOFIA; see www.sofia.usra.edu). SOFIA is a converted Boeing 747SP aircraft equipped with a 2.7-m telescope. The observatory (shown in Figure 4.29) includes several optical spectrometers for observations at wavelengths between 4 µm and 210 µm. At an operating altitude of about 13 km, SOFIA flies above most of the IR-absorbing and -emitting atmosphere. However, because of the remaining atmospheric layers and the relatively high temperature of SOFIA's mirrors (about 240 K), the SOFIA instruments are less sensitive than those of Spitzer and Herschel by factors between about 10 and 10^2. On the other hand, compared with the space observatories mentioned earlier, SOFIA has the great advantage of a much longer expected lifetime (20 years or more) and the possibility of upgrading all onboard instruments to current technological standards. Although the costs for developing and launching an airborne observatory are lower than those for space telescopes, the costs of operating an airplane observatory are significant.

The SOFIA spectrometer EXES was described in Section 4.4.7. In addition to EXES, the SOFIA observatory carries, as a first-generation instrument, the grating spectrometer FIFILS with two Ge:Ga semiconductor arrays as detectors. FIFILS operates in the wavelength range 42 µm to 210 µm with spectral resolutions up to about $R = 3,700$. (For details see Klein et al., 2010.)

Two other SOFIA spectrometers, GREAT and SAFIRE, are heterodyne instruments. They will be discussed in Chapter 9.

5

Other Techniques for the Optical Spectral Range

Although the grating instruments discussed in the preceding chapter dominate present-day astronomical spectroscopy at optical wavelengths, currently there are several other techniques in use at ground-based and space-based observatories. This chapter outlines the principles, the present applications, and the potential of three of these alternative methods.

5.1 Fabry-Perot Techniques

Fabry-Perot (FP) devices are based on the interference of light rays that are multiply reflected between partially transmissive mirrors. In astronomy, the main applications of the FP technique are special spectrometers and narrow-band interference filters. FP spectroscopy was developed in the final years of the nineteenth century by the French physicists Charles Fabry and Alfred Perot. The technique was soon applied to astronomy. Alfred Perot[1] himself used this method for measuring motions in the solar atmosphere.

The basic principle of the FP technique is outlined in Figure 5.1. The heart of any FP device is a pair of partially transmissive mirrors, separated by the distance l. In the context of FP devices, such a mirror pair is called an *etalon* (from the French designation of a standard length). In Figure 5.1, the light rays from the telescope are assumed to enter the space between the mirrors at angles γ to the mirror normals from the left (A). As indicated in the figure, the partially transmissive mirrors result in multiple reflections at the two mirror planes. At each reflection a fraction of the light passes through the mirror, while another fraction is returned into the space between the mirrors. The overall effect of the mirror pair on incident light is determined by the interference of the direct and

---

[1] Sometimes (incorrectly) spelled "Pérot."

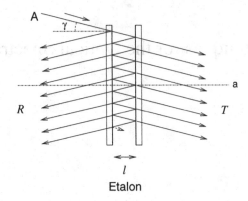

Figure 5.1. Light rays in a Fabry-Perot etalon.

of the multiply reflected rays. The effective transmission (and reflection) can be derived by calculating the field strength and phase of all waves and adding up all their contributions at the mirror planes. The corresponding calculation can be found in all textbooks of optics. Therefore, only the main results and their applications will be summarized here. Moreover, only the case with negligible absorption and ideal reflection will be discussed, and refraction effects will be neglected. In particular, we assume that the phase shift at the reflections between the mirrors is exactly π. In this case, the sum of the reflection phase shifts of all rays leaving the etalon in the forward direction is zero or 2π, as these rays experience either no or an even number of reflections. Thus, the reflection phase shifts have no effect on the interference of these rays and can be neglected.

With the assumptions made above, one gets for the effective transmission T_e of an etalon the relation

$$T_e = \frac{(1-r)^2}{1+r^2 - 2r\cos(4\pi n l\lambda^{-1}\cos\gamma)}, \tag{5.1}$$

where r is the reflectivity of the mirror surfaces, l is the distance between the mirrors, and n is the refractive index of the medium between the mirrors. Making use of the relation $2\sin^2(x/2) = 1 - \cos x$, Equation 5.1 can also be written as

$$T_e = \frac{1}{1 + F\sin^2(2\pi n l\lambda^{-1}\cos\gamma)}, \tag{5.2}$$

where

$$F = \frac{4r}{(1-r)^2}. \tag{5.3}$$

These relations can also be used to calculate the effective reflection R_e of the etalon, as in the absence of absorption we have $R_e = 1 - T_e$.

According to Equation 5.3, for a given wavelength λ the transmission T_e has maxima when

$$2n\,l\cos\gamma = m\lambda, \tag{5.4}$$

where m is an integer. This result is readily understood, as $2n\,l\cos\gamma$ is the optical path length of a ray that passes twice the space between the mirrors. At the wavelengths corresponding to Equation 5.4, a ray originating at one of the mirrors and reflected back by the other mirror has the same phase again at the plane of origin. Thus, the condition in Equation 5.4 results in constructive interference.

More surprising may appear the fact that (independently of r) for the wavelengths meeting the condition in Equation 5.4, the effective transmission is exactly $T_e = 1$. Thus, the transmission is the same as if the etalon were absent, even if the two mirrors taken alone are highly reflective. Obviously, because of the interference of the rays and the resulting oscillating electromagnetic field between the mirrors, the behavior of an etalon is quite different from that of the individual mirrors.

According to Equation 5.1, T_e varies periodically when λ, l, n, or γ is modified. If the other parameters are kept fixed, maxima occur at wavelengths that differ by $\lambda_m - \lambda_{m+1}$. This wavelength difference of consecutive maxima is called the the *free spectral range* (FSR).

From Equation 5.4 we get for the FSR

$$\text{FSR} = \lambda_m - \lambda_{m+1} = \frac{\lambda^2}{2n\,l\cos\gamma + \lambda} \approx \frac{\lambda^2}{2n\,l\cos\gamma} = \lambda/m, \tag{5.5}$$

where we assumed $m \gg 1$.

From the expressions for T_e it also follows that the effective transmission has minima (and the effective reflection R_e has maxima) when

$$2n\,l\cos\gamma = (2m+1)\lambda/2. \tag{5.6}$$

Whereas the value of the maximal transmission $T_e = 1$ is independent of r, the maximal reflection (and minimal transmission) depend on r. From Equation 5.2 we get for this dependence

$$R_{max} = 1 - T_{min} = \frac{4r}{(1+r)^2}. \tag{5.7}$$

Figure 5.2 presents an example for a FP transmission curve as calculated from Equation 5.1. As shown by the figure, the transmission shows sharp maxima

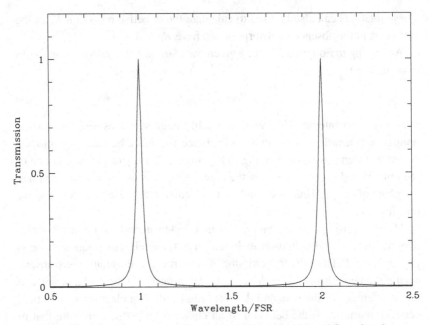

Figure 5.2. Effective tranmission T_e of an ideal FP etalon with $r = 0.9$ as a function of the wavelength. The wavelength is given in units of the free spectral range.

and broad minima. The FWHM of the maxima $w_{1/2}$ is often expressed by the relation

$$w_{1/2} = \frac{\text{FSR}}{\mathcal{F}}, \tag{5.8}$$

where for an ideal etalon

$$\mathcal{F} \approx \frac{\pi F^{1/2}}{2} = \frac{\pi r^{1/2}}{1 - r}. \tag{5.9}$$

The function $\mathcal{F}$ is called the *finesse* of the etalon (*finesse* being a French word for narrowness). From this relation we get for the example of Figure 5.2 (with $r = 0.9$) a finesse of about 30. For an ideal etalon, the finesse depends only on r. It becomes larger (as the maxima become narrower) with increasing r. This is due to an increase of the effective number of interfering rays.

Combining Equations 5.4, 5.5, and 5.8, we get for the spectral resolution of an FP etalon

$$R = \frac{\lambda}{\Delta \lambda} = \frac{\lambda}{w_{1/2}} = \frac{\lambda \mathcal{F}}{\text{FSR}} = \lambda^{-1} \mathcal{F} 2n \, l \cos \gamma = \mathcal{F} m. \tag{5.10}$$

As expected, the spectral resolution increases with the order number m and with the finesse. By comparing Equation 5.10 with Equation 3.18, we see that

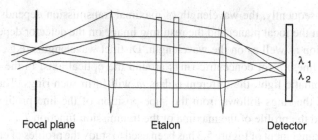

Focal plane   Etalon   Detector

Figure 5.3. Schematic layout of a simple Fabry-Perot spectrometer.

relative to the corresponding equation for interfering beams of equal intensity the number of rays N is replaced by the finesse.

According to Equation 5.10, a high spectral resolution can be achieved either by means of a high finesse (i.e., high values of r) or by a large mirror distance l. Although etalons are commercially available that have mirror reflections $r > 0.97$ (corresponding to a theoretical finesse > 100), the effective finesse of astronomical FP etalons rarely is larger than about 50 for visual applications and about 100 for IR devices. Higher numbers would require correspondingly large numbers of effective reflections. As a result of the many reflections, even small surface irregularities of the mirrors eventually add up to significant phase errors, which reduce the effective finesse. To reach a finesse of 50, a surface accuracy of the mirrors of at least $\lambda/100$ is required. Therefore, it is much easier to reach a high spectral resolution by increasing m, which in practice means increasing l. In this way, resolutions matching any line widths occurring naturally in astrophysics can be reached with FP devices. However, according to Equation 5.5, with the order m the free spectral range decreases $\propto m^{-1}$. Therefore, high-resolution FP instruments are typically used in small spectral ranges – for example, by studying profiles of single lines. Because of the small free spectral range, high-resolution FP devices must use filters to avoid order overlaps. Alternatively, a sequence of two or more etalons with different values of l can be used. In this way, unwanted orders can be efficiently suppressed, but the overall throughput of such a "polyetalon FPS" is significantly lower. Nevertheless, such devices are the best choice for high-resolution absorption line FP spectroscopy.

Fabry-Perot etalons can be applied to astronomical spectroscopy in several ways. A simple example is given in Figure 5.3. As in the case of a grating spectrometer, the arrangement consists of a collimator and a camera and the FP etalon is placed in the parallel beam. Different positions in the focal plane of the telescope correspond to different angles γ at the etalon. Thus, light rays originating from different points in the focal plane have different values of γ. Consequently, the wavelength of maximal transmission depends on the position in the focal plane, and the resulting image on the detector depends on this position as well as on the wavelength. Distinct wavelengths (e.g., narrow emission lines) form concentric rings around the optical axis. In the case of monochromatic light, the different orders m will form such rings. The radial profile of the rings follows from the superposition of the line profile of the source and the profile of the maxima of the tranmission function T_e.

The arrangement of Figure 5.3 has been used to study the profiles of emission lines of homogeneous extended objects, such as galactic nebulae. For more complex objects, however, the mix of spatial and wavelength information on the detector makes this arrangement inconvenient. Therefore, most modern astronomical FP spectrometers use only a single order m that corresponds to $\gamma = 0$. This order is separated by means of a sufficiently small circular focal plane stop. In this case, for a fixed value of $l\,n$, only one wavelength is imaged on the detector. To obtain a spectrum, the optical path length $l\,n$ is varied. Usually this is done by changing the distance between the mirrors. Because it is very important to keep the mirrors exactly parallel, accurate piezoelectric or magnetic actuators are used for this purpose. From Equation 5.2 it is clear that the wavelength of maximal transmission can also be scanned by keeping l constant and modifying the refractive index n of the medium between the mirrors. For this purpose, the etalon is enclosed in a pressure vessel and the air (or other gas) pressure is modified, making use of the fact that in a gas, the refractive index n varies with the density. Examples are the polyetalon pressure-scanned PEPSIOS devices (see Mack et al., 1963). In this technique, the parallelism of the mirrors is automatically well conserved, but reproducible wavelength scans are more difficult.

FP spectrometers in which the optical path length is modified work like tunable filters with a very narrow band width. Each wavelength scan results in a sequence of quasi-monochromatic images. Because the cosine of a small angle is close to 1, small angles γ result in only minor changes of the transmission wavelength of an etalon (see Equation 5.4). Therefore, the use of a tunable etalon is not restricted to point sources, but moderately extended objects can be observed without a significant loss of spectral resolution. Because of the low sensitivity to the angle of incidence (relative to gratings), FP etalons are often used directly in the convergent beam of a telescope (as in the configuration shown in Figure 5.4). Because of the overlap of the convergent beams toward different points in the focal plane, placing an etalon in the convergent beam reduces the dependence of the transmission wavelength on the position in the field. On the other hand, the finite range of angles γ in the convergent beam results in a lower spectral resolution. Thus, depending on the application, a

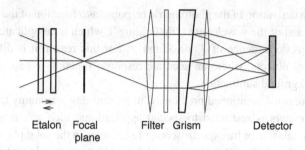

Etalon   Focal   Filter Grism   Detector
         plane

Figure 5.4. Schematic layout of a cross-dispersed Fabry-Perot spectrometer.

compromise between spectral resolution and wavelength uncertainty must be found during the design of an FP system.

A good description of the details of a modern FP spectrometer (the GHαFaS instrument of the William Herschel Telescope on La Palma) has been given by Hernandez et al. (2008). This article also describes many practical details, which for reasons of space cannot be included here.

FP techniques can also be used in cross-dispersed spectrometers. An example is the layout given in Figure 5.4. In this example, a grism is used as the cross disperser and an FP etalon as the main disperser. As in grating spectrometers, the cross disperser grism separates the different orders of the etalon. In this way, separate quasi-monochromatic images with wavelengths corresponding to different orders m of the etalon can be produced simultaneously. By tuning the etalon, spectral sequences of such narrow-band images can be obtained. An FP spectrometer operating according to this principle is the GriF NIR instrument at the CFHT telescope on Mauna Kea, Hawaii (see Clénet et al., 2002).

Fabry-Perot spectrometers make it possible to obtain high spectral resolutions with relatively simple optical systems. Because of the weak dependence of the effective wavelength on the angle of incidence, relatively large focal-plane stops can be used and spectral images of extended objects can be obtained. Thus, FP spectrometers provide a natural alternative to the IFS instruments described in Section 4.7. A disadvantage of the FP instruments is the small free spectral range at high resolution. More serious is the fact that during the spectral scans with a tunable etalon at given instant only a narrow spectral range is recorded (while the other wavelengths are reflected by the etalon). Thus, different wavelengths are recorded sequentially. This makes FP spectrometers less efficient than those using the IFS techniques described in Section 4.7. For this reason, FP spectrometers are competitive only for studies of small spectral ranges, such as the derivation of line profiles and of radial velocity maps based on single or only a few spectral lines. A practical problem of FP spectroscopy is the accurate derivation of the photometric response as a function of the position in the field and of the wavelength ("flatfielding"), which is complicated by the simultaneous dependence of T_e on λ and γ. For this reason it is difficult to observe faint objects, where an accurate response function is required for a correct background subtraction.

For the reasons mentioned previously, in present-day astronomy the use of FP spectrometers is restricted to special applications, such as emission-line mapping of galaxies or line spectroscopy of structures in the solar photosphere. On the other hand, the FP concept is widely used in astronomical photometry in the form of interference filters. All interference filters are based on thin etalons (or a combination of etalons) that consist of partially transparent mirrors separated by a transparent optical material. Although they operate at lower orders and are not tunable, these etalons have the same basic properties as those discussed earlier. In particular, as explained previously, the wavelength of maximal transmittance depends on the angle of incidence γ. Therefore, if they are used in a parallel beam behind a collimator, the effective wavelength of interference filters depends on the target position in the field of view. This complicates the analysis of images obtained with interference filters. As in the case of normal FP etalons, this field effect is less pronounced if the filters are inserted into the convergent beam of a telescope or a camera. However (for the reasons described above), in this case the bandwidth of the filter is broadened.

5.2 Fourier Transform Spectrometers

Fourier transform (FT) spectrometers make use of the fact that image information can be presented in two different ways. First, it can be characterized by a function $I(x, y)$, which gives the intensity or the flux as a function of the coordinates x and y in the image plane. Alternatively, the same information can be provided by a function $F(u, v)$, where u and v are spatial frequencies or spatial wavenumbers. The functions I and F are Fourier pairs. Each of these functions can be converted into the other one by means of the mathematical algorithm called a *Fourier transform*. In the case of a one-dimensional image function (such as a spectrum), the corresponding transformation can be written as

$$F(u) = \int_{-\infty}^{\infty} I(x) e^{-2i\pi ux} dx, \qquad (5.11)$$

where $e^x = \cos x + i \sin x$. Whereas the image function I is always real, F can be complex. If I is symmetric relative to $x = 0$, the symmetric cosine functions suffice to represent the image. Thus, in this case, F is real.

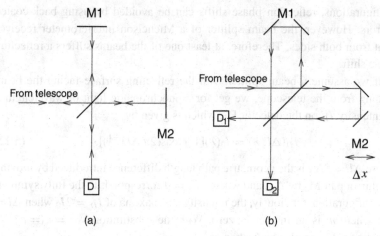

Figure 5.5. Two variants of the Michelson interferometers. M1 and M2 are mirrors. D stands for "detector."

Most FT spectrometers (FTs) use a Michelson interferometer to derive the Fourier transform of a spectrum. Figure 5.5 shows schematically two variants of this type of interferometer. In both cases, the light from the telescope is split into two beams of equal intensity, which are recombined by passing the beam splitter once more. In the simpler variant (a), part of the light is sent back to the direction of the source (and lost) when the beams pass the beam splitter the second time. In astronomy, light sources are faint and losses must be avoided. Therefore, normally the variant (b) (with two outputs) is used. Sometimes the Michelson interferometer is replaced by a *Mach-Zehndner interferometer*. The Mach-Zehndner interferometer has similar properties as variant (b) of Figure 5.5, but it uses two different beam splitters to separate and to recombine the beams.

Depending on the optical path lengths of the two arms of the interferometer, the recombination of the beams in Figure 5.5 can result in constructive or destructive interference. The light intensity at the two detectors can again be calculated by adding up and squaring the electric fields of the interfering beams. Phase differences can be produced by moving one of the mirrors (or mirror pairs) in the beam direction by a distance Δx. In addition to geometric path length differences, possible reflection phase shifts at the mirrors must be taken into account. This applies in particular to the beam splitter, which normally is a plane-parallel glass plate with a thin, semireflective metal layer on one surface. Optical theory shows that the reflected beam gets a phase shift of π, when the light enters the reflective layer from air (or vacuum), whereas there is no phase shift when the beam enters from inside the glass. In many optical configurations, reflection phase shifts can be avoided by using back-coated mirrors. However, the beam splitter of a Michelson interferometer receives light from both sides. Therefore, at least one of the beams suffers a reflection phase shift.

If we assume a beam splitter with the reflecting surface facing the beam coming from the telescope, we get for monochromatic light of wavelength λ an intensity I_D on the detector D_1, which is given by

$$I_D(\Delta l, \lambda) = I_0(\lambda)[1 + \cos(2\pi \Delta l \lambda^{-1})], \qquad (5.12)$$

where $\Delta l = 2\Delta x$ is the geometric path length difference introduced by moving the mirror pair M_2 by Δx, and where $\Delta x = 0$ corresponds to the fully symmetric configuration. Obviously, the intensity has maxima of $I_D = 2I_0$ when $\Delta l = m\lambda$, where m is an integer or zero. With the substitution $\lambda^{-1} = \kappa(= c^{-1}\nu)$, Equation 5.12 can also be written as

$$I_D(\Delta l, \kappa) = I_0(\kappa)[1 + \cos(2\pi \Delta l \kappa)]. \qquad (5.13)$$

The intensity observed for non-monochromatic light can be derived by integrating over all wavelengths (or all values of κ). Thus, we get

$$I_D(\Delta l) = \int_0^\infty I_D(\Delta l, \kappa) d\kappa = 0.5 I_D(0) + \int_0^\infty S(\kappa) \cos(2\pi \Delta l \kappa) d\kappa, \qquad (5.14)$$

where $I_D(0)$ is the total intensity observed for $\Delta l = 0$, and $S(\kappa) = I_0(\kappa)$ is (apart from numerical factors) the spectrum of the incident light. The functions $I_D(\Delta l)$ and $I_D(0)$ can be determined by a precisely measured movement of either one of the mirrors M1 or M2 of the Michelson interferometer. The last term in Equation 5.14 obviously represents a (real) Fourier transform. Therefore, the spectrum of the incident light can be calculated from the measured functions according to

$$S(\kappa) = \int_0^\infty [I_D(\Delta l) - 0.5 I_D(0)] \cos(2\pi \Delta l \kappa) d(\Delta l). \qquad (5.15)$$

To measure a spectrum with an FTS, one of the mirrors or mirror pairs (e.g., M2) is moved in the beam direction and the $I_D(\Delta l)$ is recorded as a function of $\Delta l = 2\Delta x$. From the Fourier theory, it is clear that small changes of Δx result in information on the low spatial frequencies of the spectrum, whereas large movements result in information on the high spatial frequencies. Obviously, in a real instrument the range of movements is finite. Therefore, at first glance the infinite integration range in Equation 5.15 appears to cause a problem. In practice, this is not the case, as the intrinsic spectral resolution of astronomical spectra is always finite. Therefore, for sufficiently high values of Δl, the function I_D becomes zero, and accurate results can be obtained with a finite range of mirror movements. Moreover, the effect of a finite integration on the spectrum can be calculated and corrected numerically.

So far we discussed the intensity at the detector D_1 only. When calculating the intensity at the detector D_2, we must take into account that (with the preceding assumptions) one of the beams reaching the detector D_2 suffers an odd number of reflection phase shifts (whereas for the beams to D_1, the number is always even). Therefore, for monochromatic light at D_2, the intensity becomes

$$I_D(\Delta l, \lambda) = I_0(\lambda)[1 - \cos(2\pi \Delta l \lambda^{-1})]. \tag{5.16}$$

Integrating Equation 5.16 over all wavelengths again results in an expression that contains a Fourier transform of the spectrum, and the spectrum of the incident light can again be obtained from the intensity on the detector by a Fourier transformation. Because the spectra obtained from the two detectors should be identical, they can be added and their difference can be used to estimate errors.

Compared with gratings and the FP technique, FTSs use (for each detector) only $N = 2$ interfering beams. Thus, according to Equation 3.18, the spectral resolution of an FTS is $R = 2m = 2\Delta l/\lambda$. Although N is small, a high spectral resolution can be achieved by moving M2 over a large distance. The spectral resolution can be adjusted continuously by selecting the scan range of Δx. Another advantage of an FTS is that it comprises only few and relatively simple optical components, which do not need to be as precise as they do in the case of FP instruments. Because of the simple optical layout, FT instruments are less affected by stray light than other types of spectrometers. Finally, for the same reason as was explained for Fabry-Perot devices, in an FTS, a large focal plane aperture has much less influence on the spectral resolution than in the case of a grating spectrometer. This property also makes it possible to design imaging FTSs, in which imaging arrays are used as detectors.

The main disadvantage of the FT technique is that information on different spatial frequencies of the spectrum must be recorded sequentially by scanning through a range of Δl values. If an observation is photon-noise limited, for a given spectral resolution a scanning FT device has about the same efficiency as a single-pixel (wavelength) scanning grating spectrometer. However, current grating spectrometers use imaging detectors, which record all spectral information simultaneously. Therefore, if the accuracy of a spectral observation is limited by the photon statistics, for the same S/N the integration times at an FT spectrometer can be larger by factors up to 10^5. On the other hand, at long optical wavelengths, at which diffraction patterns restrict array detectors to relatively few pixels, and where detector noise is more important than photon noise, the FTS can have significant advantages. If detector noise determines the S/N of a spectral measurement, the FT technique gains relative to grating instruments, because in an FTS the total noise of the signal and that of the detector are always present at the detector. In the case of a (single-pixel) scanning grating spectrometer at each spectral element, the full detector noise, but only the signal of the corresponding wavelength interval, is present. Therefore, as pointed out first by P. B. Fellgett (1949), if detector noise limits the observational accuracy, and if single-pixel detectors are used, FTSs can achieve higher S/N values. In the literature, this effect is referred to as the *Fellgett advantage*.

Because of the advantages and drawbacks of the FT technique outlined earlier, in astronomy FTSs have always been restricted to very special applications, in which they play an important role, however. Among these areas is the spectroscopy of extended bright sources at visual and IR wavelengths. An example is accurate wavelength measurements in the spectrum of the solar disk, for which FTSs have replaced the Fabry-Perot devices. Another traditional application of FTSs has been the measurement of accurate energy distributions of bright stars. Apart from their lower stray light effects, for this application the FTSs have the advantage of permitting the use of large focal plane apertures, which avoid photometric errors due to slit effects. Currently, FTSs are most popular at MIR and FIR wavelengths, where the S/N is limited by detector noise and where use of the Fellgett advantage can be made. Notable recent examples are the imaging FTSs on board the Japanese IR satellite AKARI (see Kawada et al., 2008), which combined an FTS with Ge:Ga semiconductor arrays; and the SPIRE instrument on the Herschel space observatory, which combines an imaging FT spectrometer covering the wavelength range 194 to 671 μm with bolometer arrays. For more details on this instrument (and references to similar devices) see Griffin et al. (2010) and Chapter 9 of this book.

Because of their low stray light levels and the very direct relation between the measured values Δx and the observed wavelengths, FTSs are also valuable as calibration devices for other types of spectrometers.

5.3 Direct Detection of Visual Photon Energies

The spectrometers that have been discussed so far use interference or refraction effects to sort photons of different wavelengths before they are recorded by a detector. From the standpoint of basic physics, this is an awkward roundabout, as (because of the unambiguous dependence of the photon energy on its frequency, according to $E_{phot} = h\nu$) each photon already contains spectral information. As pointed out in Chapter 1, detectors that directly measure the energy (and thus the frequency and wavelength) of photons are routinely used in X-ray and gamma astronomy. Examples are the X-ray CCDs. These detectors are based on the same principle as the CCD used at optical wavelengths, but they can measure the energies of individual keV X-ray photons, as the corresponding $h\nu$ values are typically 10^3 times larger than the band gap of Si (which is 1.12 eV). Therefore, each X-ray photon can produce on the order of 10^3 photoelectrons. Because the number of the photoelectrons, and consequently the electric charge produced by a single X-ray photon, is about proportional to the photon energy, the observed charge pulses give the photon energy directly.

The energy of a visual photon (2 eV at $\lambda = 620$ nm) is of the same order as the band gap of silicon (Si). Thus, Si CCDs easily detect visual photons, but they cannot measure their energy. Semiconductor detectors with smaller band gaps are available and are regularly used in IR astronomy. Examples are the impurity band detectors, which absorb IR photons by lifting electrons from an energy level above the valence band into the conduction band. However, in these materials, visual photons are still absorbed with a higher cross section from the valence band. Thus, impurity band devices cannot be used as energy-sensitive detectors for visual photons. On the other hand, the principle feasibility of energy-sensitive photon absorption in the visible range is convincingly demonstrated by the human eye. During the past decades, several types of detectors have been developed that, at least in principle, make it possible to detect not only the number, but also the energy, of single visual photons. In the following sections, the basic concepts and the scientific potential of two such techniques are described. In both these cases, the practical feasibility has already been demonstrated by actual astronomical observations. Some additional methods, which are still experimental at present, will be reported in the context of Chapter 10.

5.3.1 Microbolometers

A conceptually simple method of deriving the energy of photons is to measure the internal energy change that its absorption causes in an absorber. If the photon is completely absorbed, the energy difference obviously must be equal to the original photon energy. Because the internal energy of a solid is given by its temperature, this requires a measurement of a temperature change. Devices that measure incident radiation by determining the resulting temperature change are called *bolometers*. Bolometers have a long history. In fact, bolometers were the first photoelectric detectors employed for astronomical observations (by S. P. Langley in 1880). In present-day astronomy, bolometers are used predominantly in submillimeter astronomy. Therefore, they will be discussed in more detail in Chapter 9.

To produce a measurable temperature change in a solid by the absorption of a single visual photon with a typical energy of 2 eV (corresponding to 3.20×10^{-19} J), the absorber obviously must have a very small heat capacity. Because the heat capacity depends on the mass, very small *microbolometers*[2] have been developed for this purpose. In addition to a small mass, microbolometers must be highly sensitive to even minor temperature changes. Both conditions are met by miniaturized superconducting *transition edge sensors* (TESs). In a superconducting material (usually a metal), the electric resistance becomes zero if the temperature drops below a certain critical value. A TES is a superconductor that is operated just below this critical temperature, which means just at "the edge" of a transition to the (lower) normal conductivity. Thus, any energy input resulting in a temperature increase leads to a transition to the normal state and to a dramatic change in any current through the device. At present, TES technology is used in various fields of astronomy, as well as in particle physics.

TES-based energy-resolving photon detectors have been used successfully in X-ray astronomy for several years. Detectors of this type for visual wavelengths have been developed during the past decade mainly by a group at Stanford University in California. These devices consist of a thin (40-nm) superconducting tungsten film operated at a temperature of 100 mK. A photon absorbed by the tungsten film results in a short temperature pulse. The resulting conductivity change is measured using SQUID electronics. An integration over the resulting current pulse gives the energy of the observed photon. Apart from laboratory studies, some astronomical observations have been made, using single-pixel devices as well as a 6×6 pixel array. The maximum spectral resolution has been about $\lambda/\Delta\lambda = 15$ at 500 nm, but the quantum efficiency has been only about QE $= 0.5$ or lower. Being bolometers, these devices can, in principle, be used over a wide wavelength range, from the UV to the IR. Thus, for some applications, TES microbolometers could become competitive with semiconductor arrays if their pixel formats and quantum efficiencies can be increased, and if these detectors and their very-low-temperature operating systems can be manufactured at acceptable costs. Many technical details of these visual-wavelength TES microbolometers, first experiences, and some scientific results have been described by Burney (2007) and Romani et al. (2003).

---

[2] In analogy to particle detectors that completely absorb particle beams, these devices are also called *microcalorimeters*.

5.3.2 Superconducting Tunnel Junctions

An alternative technology for the direct measurement of visual photon energies is provided by *superconducting tunnel junctions* (STJs). As in the case of the TES devices, in STJs photons are absorbed by a superconducting metal. However, whereas TES microbolometers make use of the temperature increase caused by the absorption of a photon, STJs exploit directly the interaction of photons with electrons in a superconducting solid.

The properties of STJs result from the behavior of the electrons in metals at very low temperatures. Because of quantum effects, the electrons in a solid normally have a characteristic energy distribution, which is known as *Fermi distribution*. An important parameter of the Fermi distribution is the *Fermi level*, which is defined by the condition that the probability of finding an electron with an energy above or below this level is just 0.5. In a metal, the Fermi level is within the conduction energy band (see Section 1.3), in which electrons can gain energy and move around to produce an electric current. In a semiconductor or insulator, the Fermi level is located in the *band gap* between the valence energy band and the conduction band, which was defined in Section 1.3. As explained in Section 1.3 (and illustrated in Figure 1.6), the absorption of a photon can lift a valence band electron into the conduction band. The resulting charge in the conduction band can then be used to record the absorption event.

In a metal, photon detection according to this scheme is normally not possible, as the conduction band is always populated. Consequently, photon absorption has no significant influence on conductivity.

In a normal metal, the Fermi level has the additional property of forming the upper limit of electron energies if the metal is cooled to zero K. However, as explained in detail in physics textbooks, superconducting metals, which are cooled to a temperature below their critical temperature, behave somewhat differently. In these solids, below the critical temperature, part of the electrons form pairs, which are called *Cooper pairs*. The electrons of Cooper pairs have antiparallel spins. Therefore, Cooper pairs have zero total spin, behave like bosons, and condense to a single energy state. As a result, an energy gap δE forms between the maximum electron energy at zero K and the Fermi level. This leads to the formation of an energy gap of $2\delta E$ between the uppermost occupied energy level and the levels where the electrons can move freely. The material still remains conductive, as the Cooper pairs now act as charge carriers. However, because the lower energy levels are fully occupied and energy levels in the gap are excluded, the current carriers cannot lose or gain energy. The resulting electron energy distribution of a superconductor, with a gap between the occupied levels and the conduction region, obviously resembles that of a semiconductor. Therefore, superconductors can be used in a similar way as semiconductors to detect individual photons. However, in superconducting metals the absolute widths of the band gaps are only of the order 0.1 to 1 meV. Therefore, photons that are much less energetic can be detected and many conduction electrons can be generated by a single visual photon.

The superconducting tunnel junctions, which make use of the effects described previously, consist of two layers of a superconducting metal, which are separated by a thin (typically 100 nm) insulating layer. In radio astronomy, such devices are called *superconducting-insulator-superconducting junctions* (SIS). The insulating layer prevents a conventional current between the two superconducting regions, but its small extent allows electrons to pass through it by the quantum-mechanical tunnel effect. By applying a small voltage to the junction, the Fermi level and energy gap on one side of the insulating layer can be shifted slightly upward. This ensures that conduction electrons move (essentially) in one direction only, producing a directed electric current. Moreover, to make sure that an STJ works efficiently, measures (which cannot be discussed here) must be taken that avoid Cooper-pair tunneling through the insulator.

Photons with $h\nu > 2\delta E$ that are absorbed by an STJ dissociate Cooper pairs and produce about $2(h\nu)/\delta E$ conduction electrons, which can tunnel through the insulator and can result in a corresponding current pulse. If there are no losses, these pulses are proportional to the original photon energy. The energy resolution is related to the variance of the number of the conduction electrons produced by a single photon.

STJs can be either compact microdevices or they can consist of a somewhat larger superconducting absorber with small readout junction elements at their edges. The latter type of STJs are called *distributed readout imaging devices*, or DROIDs. In DROIDs, the position of the photon absorption can be determined from the relative strength of the current pulse at the individual readout junctions. Thus, in addition to the energy resolution, DROIDs have a (limited) imaging capability. The schematic layout of a DROID is given in Figure 5.6.

As noted earlier, the energy resolution of an STJ is related to the variance of the number of conduction electrons produced by the photon absorption. Assuming that the variance is proportional to the square root of the carrier number, we can expect the energy resolution to be about proportional to the square root of the energy. This turns out to be a good approximation. The exact numbers of current carriers and of their variance are influenced by a series of complex processes that take place after the absorption of a photon in the superconductor and in the tunnel layer. A discussion of these details is outside

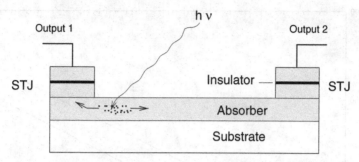

Figure 5.6. Schematic layout of DROID-type superconducting tunnel junction. Conduction electrons produced by the absorption of a photon in the superconducting absorber produce a current pulse by tunneling through the insulator. The total number of electrons is a measure of the photon energy, and the relative strength of the pulse at the two outputs provides information on the impact position of the photon.

the scope of this book. The reader interested in this topic will find a good introduction to STJs and their astronomical applications in a recent review by Martin and Verhoeve (2010). As explained in this review, the energy resolution $R = E/\Delta E$ can be predicted reasonably well by the so-called tunnel-limited resolution approximation. The spectral resolution of STJs in this approximation is given in Figure 5.7 for two frequently used materials. Because the energy gap δE (which is related to the critical temperature) of molybdenum is about 1/10 of that of tantalum, the same photons can produce more electrons in a Mo-based superconductor, which results in a correspondingly higher energy resolution. Experience with real STJs shows that at visual wavelengths, the resolutions are rather close to the values predicted by Figure 5.7.

Single-pixel STJs and STJ arrays for optical observations have been developed and used by different groups in Europe and the United States. So far, all these devices have been based on superconducting tantalum absorbers. Existing STJ array cameras have up to 120 pixels. There also exists a 3×20 element DROID array that can resolve 660 individual pixels. Visual-band spectral resolutions up to $R = 23$ have been reported. The maximum quantum efficiency of tantalum-based STJs has been predicted to be about 0.7, although that of existing devices is only about half this value. For a complete survey of the existing devices and their performance, see again Martin and Verhoeve (2010).

Compared with TES microbolometers, the STJs have the advantage of a higher operating temperature (about 300 mK) and a faster response time, which (at least in principle) should allow higher photon counting rates. Producing larger arrays may be simpler than in the case of TES. On the other hand,

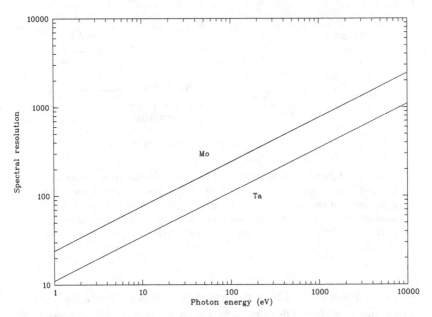

Figure 5.7. Predicted tunnel-limited spectral resolution $R = E/\Delta E = \nu/\Delta \nu = \lambda/\Delta \lambda$ of superconducting tunnel junctions made of molybdenum (Mo) and tantalum (Ta). (According to Martin and Verhoeve, 2010).

although a significant amount of work has been devoted to developing STJs (mainly at the Advanced Studies and Technology Preparation Division of the European Space Agency), so far these detectors are not yet a viable alternative to other methods of low-resolution spectroscopy or narrow-band photometry at visual wavelengths. However, the progress made during the past decade certainly justifies continuing the development of these potentially promising devices.

Closely related to the STJs are the *microwave kinetic induction detectors* (MKIDs) and the *quantum capacitance detectors* (QCDs). These photon energy-sensitive light sensors, which are being developed at various laboratories, use a different readout scheme, which may make it easier to construct large arrays for astronomical imaging and spectroscopy. The physical principles and the potential of these still-experimental devices will be discussed in Chapter 10.

6
Preparing and Reducing Optical Observations

Astronomical observing projects generally can be divided into three phases: (1) the preparation, (2) the actual observations, and (3) the reduction of the collected data.

At most large ground-based telescopes and at all space facilities, the actual observations are executed automatically by means of a dedicated computer program. The program points the telescope, configures the spectrometer, starts the individual observing steps, and stores the resulting data. The observer initiates the observations by entering the observing parameters (such as the target coordinates, the desired instrument configuration, and the integration times) into a data file, which is uploaded to the computer prior to the beginning of the observations. Only at smaller telescopes, during tests, or at seldom-used instruments are the spectrometers operated manually by the astronomer or by an operator.

All observations result in raw spectra, which, in addition to scientific information, contain artifacts caused by the instruments, by backgrounds, and (in the case of ground-based observations) by the Earth's atmosphere. Therefore, in a final (reduction) step, these artifacts must be removed, and the data must be properly calibrated.

This chapter describes mainly how spectroscopic observations are prepared and how the raw data are reduced. As the execution phase follows in a straightforward way from the preparatory work, less space is devoted to the phase of the actual observations.

6.1 Planning and Preparing Observing Runs

6.1.1 General Considerations

The planning phase is the most important part of a successful observing run. Once the observations have been started, modifications of the observing plan are often impossible. If changes can be made during the execution, they always

145

involve a loss of valuable observing time. Thus, poor planning unavoidably has negative consequences. Mistakes made during the data reduction phase are much less critical, as the reduction process can be repeated. Furthermore (assuming well-engineered and reliable instruments), problems during the actual observations are unlikely to occur if the program has been thoroughly prepared.

Any observing program starts by identifying and defining scientific objectives. The scientific aims generally result in constraints for the wavelength range, the spectral resolution R, and the signal-to-noise ratio (S/N), which are required to achieve the objectives. Because the observing time increases with R and S/N (see, e.g., Equation 3.40), these values must be chosen carefully. As pointed out in Section 2.6, a higher spectral resolution generally results in more information. However, aiming for values of R and S/N that are above those required by the scientific objective may result in unacceptably large observing times or (if the time allocation is fixed) in a reduction of the observable number of targets.

The spectral range and spectral resolution also determine the type of spectrometer that can be used. After a specific instrument has been selected, the required instrumental configuration (including parameters such as the slit widths, the choice of gratings, grating angles, etc.) must be defined.

Some scientific problems require observations of uniquely determined targets, such as a certain star, planet, or galaxy. For other programs, targets can be selected from a multitude of objects of a certain class or property (such as stars of a certain spectral type). In the latter case, the next planning step of an observing program must be the selection of suitable targets and the definition of an optimal sequence of observations. Potential targets must meet various obvious conditions, such as being visible on the night sky at an acceptable zenith distance at the time of the observations.

An important input parameter for the planning phase is the amount of time required to reach an adequate S/N for each target. Order-of-magnitude estimates of the integration times can be made using simple relations such as Equation 3.41. Much more accurate values can be derived using software packages called *exposure time calculators* (ETCs), which exist for all major astronomical spectrometers. ETCs give the required integration time as a function of the spectrometer configuration and the desired S/N, either numerically or in graphic form. Because the available parameters and the efficiencies differ between spectrometers, ETCs depend on the specific instruments. ETCs are normally provided by the team who built the spectrometer and, together with instrument handbooks, are part of the documentations made available to observers. They are also found on the observatory Web pages.

During the observations, additional time is required pointing the telescope, configuring the spectrograph, the detector readout, and calibration exposures. Thus, the total time required is always longer than the sum of the integration times of all targets. The amount of overhead depends critically on how an observatory is organized. Sometimes the ETCs also assist in deriving these overheads. Otherwise the overheads must be calculated independently from information provided in the instrument handbooks. Obviously, time can be saved by keeping the slewing distances on the sky between consecutive targets small and by using short detector readout times. However, in some detectors, a fast readout leads to a higher noise level, and a slow readout may be required to meet the S/N requirements of the program.

As noted already, normally the coordinates and observing parameters for each target are entered into the database of a computer that points the telescope and controls the spectrometer. For technical reasons, the telescope and the spectrometer actually are operated by different programs. Normally, however, there exists one single interface at which the observer can enter all relevant data. When developing new spectrometers, or when modifying observing procedures, astronomers and engineers must take into account the (sometimes complex) interplay of the different computers and programs involved.

As a rule, the input files for spectroscopic observations are produced using software tools, which are provided by the corresponding observatories. Normally for each target a fixed set of parameters must be specified. These parameters include the target coordinates, the spectrometer configuration, integration times, calibration exposures (see Section 6.3), guide-star coordinates, and data readout and storage details. In the case of long-slit spectroscopy, the orientation of the projected slit on the sky may also need to be defined. The complete data set for a target is called an *observing block* (OB). Most observatories provide standard OBs for common spectroscopic programs. These standard OBs, called *templates*, can be edited and adapted to the needs of the individual observer. If the observations must be made at a certain time (as, e.g., in the case of periodically variable objects), the start time of the block must also be given. Blocks that are not time critical are normally arranged into observing sequences that are ordered to minimize the total overhead time. During the observations, the resulting queue of OBs is then executed in a fixed order (but there may be exceptions; see Section 6.2).

6.1.2 MOS Preparation

Preparing MOS observations follows the same basic steps as single-object work. However, a few additional details must be considered. In the case of single-object spectroscopy, a small coordinate error can be corrected during the target acquisition by a slight change of the telescope position. In the case of multiobject spectroscopy, such a correction is possible only if all objects have the same offset. Therefore, the individual coordinates of MOS targets must be more accurate. To prevent light losses, the MOS coordinates must be correct within a small fraction of the slit widths or the fiber diameters.

Depending on the acquisition algorithm, an MOS observer may have to provide reference stars with coordinates that must be as accurate. Reference stars are mandatory if plug plates or masks are used, when they are needed for the exact alignment of these plates.

In principle, target and reference star coordinates can be given either as coordinates on the sky (usually referred as *world coordinates*), or coordinates in the focal plane of the telescope. From the technical description of the MOS instruments in Chapter 4, it is clear that internally, MOS spectrometers always use focal-plane coordinates. Thus, if world coordinates are given, at some stage they must be converted into focal plane coordinates. Because images produced by a telescope are projections of the sky, in first approximation the two coordinate systems are linearly related. However, in practice, the optics of all telescopes produces nonlinear distortions. Moreover, in modern telescopes with active optical systems, the focal ratio (and other optical parameters) may not be exactly constant. A conversion of world coordinates to focal plane coordinates with the accuracy noted earlier is not trivial, and conversion codes are not available for all MOS spectrometers. In the absence of such conversion software, focal-plane coordinates must be determined directly by imaging the field in question with the same telescope.

In addition to distortions due to the telescope optics, image distortions due to differential atmospheric dispersion may be present in an MOS field. Obviously, this type of distortion is a function of the zenith distance and does not depend on the telescope. The effect can be eliminated by carrying out the the imaging and the spectroscopic observations at a similar zenith distance.

In the case of multifiber instruments with robotic fiber positioners, the coordinates must be uploaded directly to the spectrometer. If plug plates or masks are used, the coordinates must be converted into a standard format that can be read by computer-controlled tools (such as the Gerber format) and entered into a plate drilling or mask cutting device. This must be done sufficiently in advance of the observations to allow for the production and the installation times of the plug plates or masks.

In the case of multislit instruments, the positions and width of the projected slits must be specified. For slit masks, in addition, individual slit heights and slit tilt angles can be requested. Normally, the selection of MOS slit parameters

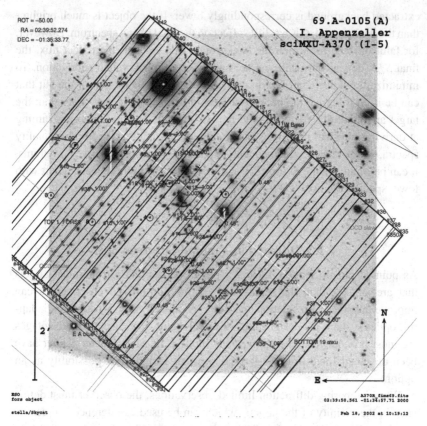

Figure 6.1. Screen shot of an interactive program that can be used to define the slit positions, slit widths, slit lengths, and slit orientations for multislit spectroscopy. In this example, 48 slits for the ESO FORS2 spectrometer (in mask mode) were defined. The targets are faint galaxies behind a relatively dense galaxy cluster. To get a clean background in the relatively short slits in spite of the high foreground density, most of the slits are tilted relative to the dispersion direction.

is carried out using interactive graphic software, with which the intended slits can be overlaid onto sky images containing the planned targets (see Figure 6.1). If these images have been taken with the same telescope and at similar zenith distances, they can be used directly. Otherwise, again, corrections for image distortions must be applied.

The optimal choice of slit heights depends on the scientific objective and the object's brightness relative to the sky background. With small slit heights, more objects can be observed in one frame. However, the region of the slit that is available for the background derivation is smaller, and the accuracy of the extracted background is correspondingly lower. If the object is much brighter than the background, this has little effect on the final object spectrum. However, for faint objects, for which the background flux exceeds the object flux, the final S/N depends critically on the quality of the background subtraction. To minimize the noise contribution of the background, the region of the slit that can be used to derive the background should be several times larger than the target area, and this region should be free of other objects. Programs aiming at deriving physical properties of the targets normally require higher quality spectra, which must be obtained with relatively long slits. For survey programs, it can be of advantage to use shorter slits to observe more objects, albeit with lower spectral quality.

6.1.3 Diffraction-Limited Spectroscopy

As pointed out in Section 4.4.1, for point sources and astronomical targets that are smaller than the seeing disk, diffraction-limited observations can improve the spectral resolution and the S/N significantly. Most current adaptive optics systems are designed for IR wavelengths, for which their advantages are most pronounced. Therefore, many modern IR spectrometers either have been designed for diffraction-limited work or include this possibility as an option.

When preparing diffraction-limited observations, the observer must define stars in the vicinity of the targets, which can be used as reference sources by the system's wavefront sensor. These stars must be within the isoplanatic plane of the target, and they must be sufficiently bright to give an adequate S/N in typical integration times of $\leq 10^{-2}$ s. The limiting magnitude of the reference stars depends on the type and quality of the wavefront sensor (but not on the telescope aperture). Numerical values can be found in the handbooks of the corresponding AO systems.

For targets for which sufficiently bright natural reference stars are not available, most observatories with large telescopes offer *artificial guide stars*, or *laser guide stars*, which are bright spots in the upper atmosphere produced by laser beams (see Figure 6.2). However, even in this case, the observer must specify a reference star in the field, which can be used to measure and to eliminate the "image motion" caused by the turbulent atmosphere. (Because the light producing an artifical guide star passes the atmosphere twice, this lowest-order seeing effect, also called *tip-tilt effect*, cannot be determined with the laser beam.) The tip-tilt effect can be determined with less-bright stars.

Figure 6.2. Laser beam, producing a laser guide star, emanating from the Gemini-N telescope of the Mauna Kea observatory, Hawaii. Image courtesy Gemini Observatory/AURA.

Thus, suitable stars for this purpose are generally available everywhere in the sky.

6.1.4 Ground-Based MIR and NIR Observations

Ground-based IR observations must cope with a particularly intense sky background. At the same time, the photon noise is less important, as there are more photons per unit energy. Therefore, depending on the exact wavelength, somewhat different observing strategies are used in the IR, which require modifications during the preparation (and reduction) phase.

The background is particularly strong in ground-based observations in the thermal infrared ($\lambda > 3\,\mu m$). As pointed out in Chapter 4, because of the thermal emission of telescopes and of the atmosphere, sensitive spectroscopy in the mid-infrared has to be carried out with cold telescopes from space. However, for some scientific objectives, the relatively small cold space telescopes do not provide an adequate angular resolution. Moreover, at high spectral resolution, the light-gathering power of ground-based telescopes can partially compensate for the higher background. Therefore, some MIR spectroscopy is carried out from the ground. In this case, special background subtraction techniques are used. These techniques can also improve the results obtained with the relatively warm passively cooled space instruments and with NIR spectrometers.

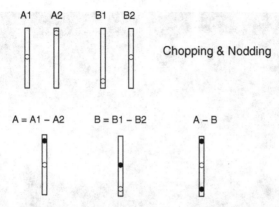

Figure 6.3. Schematic presentation of the chopping and nodding technique described in the text. Open circles show target positions in the spectrograph slit that result in positive spectra. The filled circles give the positions where (owing to the subtractions) negative spectra are formed in the final spectral frames. A spectral frame resulting from the double-subtraction procedure is shown in Figure 6.4.

As in the visual range, in IR slit spectroscopy the sky is derived and removed by measuring the background at a position in the slit outside the object. This sky flux is then subtracted from the sky-flux-contaminated target flux. In the MIR, the background (produced by the sky and the warm telescope) is so intense that only very short exposure times are possible without saturating the detector. Moreover, the MIR sky background varies rapidly with time and space. Therefore, for such observations a technique is used that is referred to as *chopping and nodding*. *Chopping* means that the field of view of the telescope alternates rapidly between two positions on the sky. Thus, the slit center changes between the target and a blank field. Chopping is achieved by oscillating the secondary mirror of a telescope in a square wave pattern with a frequency of several Hz. In spectroscopy, the chopping normally is carried out along the spectrograph slit, with the object being visible in the slit in its original or in an offset position (see Figure 6.3). Subtracting the spectra obtained in the blank field from those of the target removes the background on first order. However, the background correction is not complete, as the movement of the secondary mirror changes the optical path through the telescope. Therefore, after about 10^2 chopping cycles the telescope is pointed to a different position on the sky, where the chopping is continued. After another about 10^2 cycles, the telescope is moved back. This procedure is called *nodding*. As indicated in Figure 6.3, usually the target position and the sky position are exchanged by the nodding. Thus, at the initial telescope position (A) the target is in sky Position 1, and the reference sky in Position 2. In the nodded telescope position (B) the target is in sky Position 2, and the reference sky in Position 1. The target flux is integrated over one chopping cycle at each of the two nodding positions, A and B. If we calculate the differences indicated in Figure 6.3 (i.e, A1–A2, B1–B2, A–B), we get on the detector at the position which corresponds to the initial target position (i.e., A1 in Figure 6.3)

$$[(T + S1) - S2] - [S1 - (T + S2)] = 2T, \qquad (6.1)$$

where S1 and S2 are the background flux at the two positions and T is the target flux observed in one chop frame. Thus, at this detector position, the chopping and nodding involving four individual frames results in a spectrum with twice the flux of a single measurement. However, as can be seen from Figure 6.3, there also will be two spectra with a negative flux. The negative flux results from the fact that during one of the subtractions, the target flux has a negative sign. The absolute flux value recorded at these positions is the original flux of one measurement. Thus, in total we record four times the flux of a single frame, as is to be expected for the evaluation of four frames. Experience shows that the double subtraction removes the background much more accurately than either chopping or nodding alone. Because the many chopping cycles of an MIR observation result in a large number of spectral frames, the chopping subtraction procedure often is carried out at the telescope and only the results of nodding cycles are saved.

A disadvantage of the chopping and nodding is that integration time is lost while the secondary mirror and the telescope are in motion. Therefore, when preparing MIR spectroscopic observations, a first step must clarify whether chopping and nodding is the best choice. Whereas for MIR imaging programs chopping and nodding is always required, high-resolution spectroscopy of bright targets can sometimes be carried out more economically in the normal ("starring") spectroscopic mode. If chopping and nodding is used, the chopping direction (equal to the slit direction) and the chopping throw must be specified in a way that no objects are at the background positions. For this purpose, deep IR images must be available.

At NIR wavelengths, the background is dominated by the airglow lines. However, these lines are sharp, and between the airglow lines the NIR sky is rather dark. The noise at these wavelengths, therefore, is often given by the detector readout noise. Because the relative contribution of the readout noise increases with decreasing integration time, rapid chopping makes no sense in the NIR. However, nodding and double subtraction can be used here to remove the strong (and variable) airglow lines. An example of a two-dimensional NIR (H-band) spectrum obtained in this way is shown in Figure 6.4.

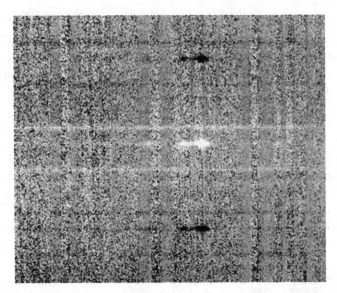

Figure 6.4. Example of an NIR (H-band) spectrum of a redshifted emission line galaxy. The (at this wavelength, particularly strong) airglow lines have been eliminated by the double-subtraction procedure described in the text. However, the positions of these lines are still visible because of the higher noise where large line flux values were subtracted. Note the presence of a central positive spectrum and two negative spectra resulting from the double subtraction. The negative spectra are visible for the main object, as well as for the two fainter continuum sources in the field. Image courtesy European Southern Observatory.

6.2 The Execution Phase

As pointed out at the beginning of this chapter, at most present-day large observatories the actual observations are normally executed automatically by means of a computer program that points the telescope, configures the spectrometer, starts the individual observing steps, and stores the resulting data. Such observations under computer control result in an optimal use of the observing time – and prevent mistakes during the execution phase. Even in the rare cases in which the individual observing steps are initiated manually by the observer or an operator, the observations normally follow a predetermined sequence. Therefore, if the observations have been well prepared, their execution often is a routine operation, which requires little additional input, and in the case of space observatories, the observer often has no possibility at all to interfere during the actual execution phase. As pointed out earlier, during such automated observations a series of observing blocks is observed sequentially. In the case of space observations, for which the observing conditions are constant or highly predictable, the observing queues can be fixed well in advance. However, at ground-based observatories, changing meteorological conditions and seeing variations can make it advisable or necessary to change observing queues during the night. Therefore, at ground-based observatories, the observations normally are still controlled or monitored by the observer or by an experienced operator, who can change the observing sequence if changing conditions make this necessary.

In the case of slit or fiber spectroscopy, the exact alignment of the targets on the slit or fiber centers is very critical. Therefore, normally a short "acquisition image" is obtained before the start of the spectroscopic exposures. This image is used to compare the actual and requested positions of the targets relative to the slits or fibers, or (if the targets are not visible on the short exposure) the actual and required positions of references stars in the field. On the basis of the observed deviations, the telescope is then moved to the correct position. This can be done either automatically by means of suitable software, or by the observer or an operator. Because for modern telescopes blind pointings deviate by not more than a few arcsec from the intended positions, these final corrections are small and very accurate. Figure 6.5 shows part of an acquisition image with the slit positions overlaid. In the case of focal-reducer spectrographs and for spectrometers in which reflection gratings can be replaced by mirrors, control images of the targets can also be obtained through the slits.

The monitoring of the execution of spectral observations can be done either at the telescope site or from a remote location where all relevant information on the status and the progress of the observation (including quick-look results) are available online. Often different interaction levels are offered to the observer. A good example is the W. M. Keck Observatory on Mauna Kea, Hawaii. Normally all observations with Keck telescopes are carried out remotely from the Keck headquarters in the town of Waimea, where the observer has full control of the spectrometers. However, observations can also be carried out from various sites on the U.S. mainland, and even from a site in Australia. There is also a mode of remote monitoring with limited interaction possibilities, or monitoring without the possibility of any interacting (for details, see the Keck home page, http://keckobservatory.org).

At space observatories, the observing conditions are more predictable and fully automatic observations are safer. However, even in space, a monitoring of the actual observations is of significant advantage, as irregularities and instrumental problems often can be detected faster by experienced personnel than by automatic procedures. Independently of the details of the observing mode, for the evaluation of the results it is important (and sometimes critical) that the observational details (such as the actual instrumental parameters, the

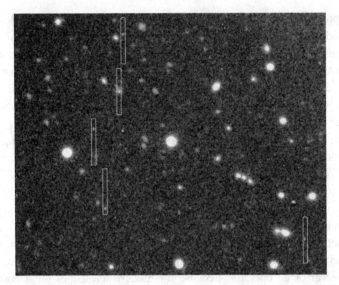

Figure 6.5. Section of a short MOS "acquisition image." Overlaid are the outlines of five spectrograph slits. The slit patterns have been generated from the information in the observing file. Such overlays can be used to align the slit pattern to the target field or to verify the alignment. The field is an area (about 2.2 arc minutes wide) at the Hubble Deep Field South. This acquisition exposure was obtained with the FORS1 instrument at the ESO VLT. Because the faint targets are difficult to see on this short exposure, their exact positions are indicated by small crosses. Image courtesy European Southern Observatory.

time line, environmental conditions, etc.) are well documented. In observations carried out under computer control, the individual steps are normally recorded in log files, and all important parameters are copied into the headers of the data files. Apart from storing and archiving the observing parameters, good observing software must enable the observer to retrieve this information quickly and in an easily readable format, if it is needed during the course of an observing run.

6.3 Calibration Procedures

6.3.1 Wavelength Calibration Strategies

As described in the preceding chapters, for spectrometers that use gratings or other interference devices, the observed wavelengths can be determined directly from the geometrical parameters of these instruments. In fact, as pointed out in Chapter 3, historically the wavelength of light was first determined in this way, and temporarily spectrometers were used to define the physical unit of length. On the other hand, direct absolute wavelength derivations require very accurate geometric measurements. If such measurements are carried out in air, they must be carefully converted to standard air conditions (which are practically never found at observatory sites) or to vacuum values. Therefore, it is much more convenient to calibrate the wavelength recorded by a spectrometer by means of standard light sources with accurately known (air or vacuum) wavelengths. In most cases, this is done by observing with the same setup spectra that contain well-defined emission lines. In the literature, the resulting spectral frames sometimes are referred to as *arcs*, because during the area of photographic spectroscopy open electric arcs were used as calibration light sources. Today the most common sources for calibration spectra are hollow cathode lamps (HCLs) or simple gas discharge tubes. HCLs are commercially available from many suppliers and with many different chemical compositions. The choice among different lamps depends on the application and on the spectral dispersion. For low-resolution work, lamps producing relatively few isolated lines (such as He-Ar-Ne discharge lamps in the visual, and Ne-Ar-Kr lamps in the NIR) are optimal, whereas high-resolution spectrometers require lamps with a large number of lines. Popular sources with a rich visual line spectrum are thorium-argon HCLs (see Figure 4.16). Th-Ar HCLs are reliable and easy to operate, but they show aging effects, and wavelength calibrations based on these lamps are sometimes impaired by the nonuniform line distribution and the highly different line intensities of the Th-Ar spectrum. Therefore, new optical wavelength calibration sources based on laser frequency combs may replace HCLs in the future (see, e.g., Araujo-Hauk et al., 2007).

The wavelength calibration spectra must be taken with exactly the same spectrograph configuration as is used for the target. If the spectrometer is stable, they can be obtained either before or after the object exposures. Often the calibration exposures are carried out during the day following the observing night. In this way, valuable observing time can be saved during the night. However, if bending or thermal effects of the spectrometer are significant or if a higher accuracy is important, the calibrations must be carried out immediately before and after the object observations. By averaging the calibration spectra taken before and after the object, linear instrumental variations can be eliminated. For very accurate wavelength measurements, the calibration spectra must be observed simultaneously with the object spectra (as in Figure 4.16).

Instead of using emission line spectra, wavelength calibration can also be carried out by observing the target through a line absorption cell. In the visual spectral range, accurate measurements have been made using iodine absorption cells. These are sealed glass vessels in which iodine is evaporated by heating iodine crystals. The resulting iodine molecules produce a large number of sharp absorption lines in the wavelength range of 500 to 610 nm. These lines are superposed on the object spectrum; wavelengths in the object spectrum can be determined by measuring positions relative to the iodine lines. A disadvantage of the iodine cells is that about 50 percent of the incident light is absorbed in these cells, which reduces the scientific usable part of the spectrum and the efficiency of the observations.

Gas absorption cells also exist for other wavelength bands. For accurate NIR spectroscopy, ammonia cells have been used successfully (e.g., Bean et al., 2010). A wavelength calibration spectrum that comes for free in the red and NIR range is the airglow emission lines of the night sky. These lines are very sharp and their wavelengths are well known. Airglow lines are also present in the visual spectral range, but their number density is too small to base a dispersion relation on the visual night sky lines. However, they are sometimes used to check and to correct the zero points of wavelength scales derived using lamps. An obvious advantage of air glow lines is that they are automatically observed simultaneously with the target spectrum. However, observers aiming at very high accuracies must take into account that the intensity of airglow lines varies with time.

Accurate wavelengths of individual spectral lines that can be used for calibration purposes can be retrieved from the Web page of the U.S. National Institute of Standards and Technology (NIST; www.nist.gov) and from the Web pages of major observatories. Lists of night sky lines that are suitable for wavelength calibration have been published, for example, by Osterbrock et al. (1996, 1997).

For calibrating Fourier-transform (and high-resolution FP) spectrometers, lasers are often used. Lasers have the advantage that their vacuum wavelengths can be derived very accurately by directly measuring their frequencies using frequency combs (see Section 3.2). At present some lasers are the most accurate frequency standards, and laser-calibrated FT spectrometers are often used to derive or to improve accurate wavelength data of atomic and molecular spectral lines.

6.3.2 Flux Standard Observations

Assuming a linear detector, the spectral signal obtained by a spectrometer is proportional to the flux from the target. However, the ratio between the signal and the incident flux depends on the efficiencies of the instrument and of the telescope, and on the detector sensitivity. All these quantities depend on the wavelength. In principle, the efficiencies of the individual components can be determined by laboratory measurements, and the total efficiency can be derived by multiplying the individual contributions. However, a reliable derivation of the efficiency of large optical components, such as telescope mirrors, is difficult, and the combination of the different contribution by multiplication tends to result in significant errors. Therefore, astronomical spectrometers normally are calibrated by observing standard stars (or planets) with a known flux and energy distribution. These standard stars have been calibrated (directly or indirectly) by a comparison with a laboratory light source of known intensity, such as a black body. Lists of suitable standard stars and their flux as a function of wavelength have been published, for example, by Oke (1990) and Hamuy et al. (1992, 1994). To exclude light losses at the slit, the flux standard stars normally are observed with a wide slit of typically 5 arcsec. Because flux standards are selected to have smooth spectra, the loss of spectral resolution due to the wide slit does not affect the calibration.

In the thermal infrared, often planets are used for flux calibration. Obviously the observed flux of these objects varies with the orbital motion of the planet and the Earth. However, these variations can be accurately predicted and corrected.

In the case of ground-based observations, the observed flux is also affected by light absorption and scattering in the Earth's atmosphere. The amount of atmospheric extinction depends on the observatory site, the atmospheric conditions, and the zenith distance. However, at good sites, there are many nights when the extinction varies only with the zenith distance and is otherwise constant in space and time. Such nights are called *photometric*. Obviously, photometric standards should be observed only on such photometric nights. Although spectroscopy of other objects may well make sense on nonphotometric nights (because relative flux spectra can already contain valuable information), observing flux standards under nonphotometric conditions serves no purpose.

6.4 Reduction of Raw Spectra

The spectra that are recorded with sensitive astronomical spectrometers normally contain, in addition to the object's signal, contributions from the sky background, the instrument itself, and the environment. As illustrated in Figures 6.6 and 6.7, in the case of faint objects the raw spectra often show little resemblance to the true spectra of the underlying targets. Therefore, before the spectra can be used for scientific purposes, the instrumental and atmospheric effects must be removed. Moreover, as the initial output signal is in arbitrary units, and because for many applications the wavelength scale resulting from the instrument design parameters is not accurate enough, an accurate

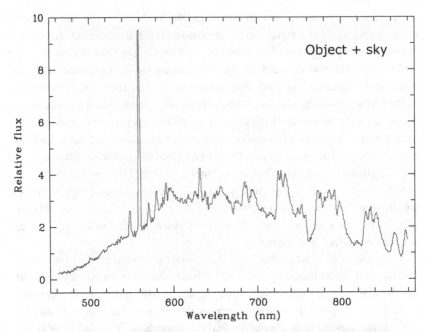

Figure 6.6. Wavelength-calibrated raw spectrum of a distant elliptical galaxy. Because (at 600 nm) the sky background is almost three times brighter than the flux from the target, the raw spectrum is strongly dominated by airglow lines, which in this low-resolution ($R \approx 350$) spectrum merge to form broad bands in the red spectral range. The nature of the underlying target spectrum is not evident from this raw spectrum.

wavelength scale must be established and the instrumental signal must be converted to physical flux units.

During the past decades, extensive software packages have been developed for the reduction of astronomical spectra. Well-known examples are the IRAF system, which was designed (and is being maintained) by the U.S. National Optical Astronomical Observatories (NOAO), and the MIDAS system of the European Southern Observatory (ESO). For many astronomical instruments there also exist software packages (called *reduction pipelines*) that permit a fully automatic data reduction. An example is the ESO-FEROS spectrometer (mentioned in Section 4.4.3), in which the reduction can be started at the telescope as soon as an observation is completed. In this case, the observer can get a fully reduced spectrum within a few minutes after the shutter has been closed. This is possible because the instrument has few observing modes and is very stable.

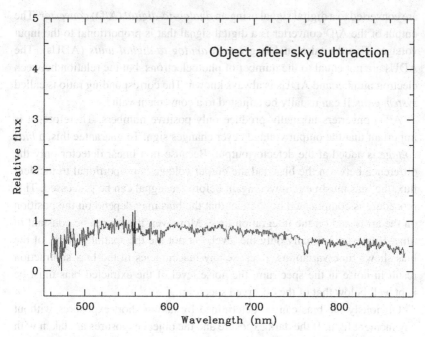

Figure 6.7. The spectrum of Figure 6.6 after subtraction of the sky background. Although the spectrum is noisy, it can be clearly identified as an elliptical galaxy with a redshift of about $z = 0.138$.

Although pipeline processing is fast and convenient, it does not always result in an optimal extraction of the information present in the raw spectra. Therefore, it is often worthwhile to repeat the reduction with parameters that give optimal results for the specific scientific objective of a program. To understand the motivation and the complex details of the reduction software and to make best use of these programs, some knowledge about the individual procedures and their physical basis is important. Therefore, in the following sections the most important spectral reduction procedures are summarized. This summary follows the scheme that is normally used to reduce spectra obtained with array detectors. However, many of these steps also apply to the reduction of raw spectra obtained with other detector types, although (as explained for some examples) they may have to be executed in a different order.

6.4.1 Bias Subtraction

The intrinsic output signals of array detector are always voltages. Normally the observer never sees these voltages, because at the output of the array the voltages are converted to a digital signal using an *analog-to-digital* (A/D) *converter*. The output of the A/D converter is a digital signal that is proportional to the input voltage. This digital signal is given in *analog-to-digital units* (ADUs). The ADUs are not equal to the number of photoelectrons, but the relation between electron number and ADUs is always known. The corresponding ratio is called *system gain*. It can usually be adjusted to a convenient value.

A/D converters normally produce only positive numbers. Therefore, it is important that the output voltage never changes sign. To guarantee this, a *bias voltage* is added at the detector output. Because in a linear detector only the difference between the bias and the output voltage is proportional to incident flux, the bias must be removed again before the signal can be processed. This procedure is complicated by the fact that the bias may depend on the position on the array and on the integration time. Moreover, it may not be constant in time. However, normally only the level, but not the the spatial pattern, of the bias shows time variations. Because any inaccuracies in the bias subtraction result in noise in the spectrum, the noise level of the extracted bias must be kept well below that of the observed spectrum.

Obviously the bias can be determined by taking short exposures without any incident light. If the dark exposure and the object exposures are taken with the same duration, such frames can also be used to remove any dark current, if present. To avoid a loss of observing time, the dark exposures normally are taken during daylight, and many dark exposures are averaged to reduce the effective noise. To take care of a possible time variability, the bias also must be measured close to or during the object exposure. For this purpose, many array detectors have "overscan sections" that are read out with the rest of the chip, but are covered and do not receive light during the object exposure. Using an average over these overscan pixels, the average bias frames can be corrected for the time variations. Because the exposure time for the overscan pixels is the same as for the target pixels, a first-order dark-current correction is achieved as a by-product.

Photon-counting detectors using photo cathodes do not have a bias. However, there may be *dark counts* that may be position dependent. Obviously, the dark count level can again be removed by dark exposures that are subtracted from the target frames.

In Section 6.1.4 techniques were described that are used in ground-based IR spectroscopy to eliminate the very high and variable IR sky background. As explained in that section, using a double subtraction scheme the background is removed efficiently as a first step of the reduction. From the description given in Section 6.1.4 it can be seen that together with the sky background the bias is automatically also removed. Thus, for IR observations that follow the

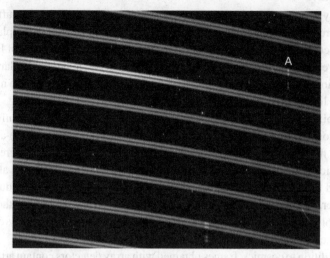

Figure 6.8. Cosmics (visible as small white spots) in an echelle spectrogram. Because the spectrum was obtained using a fiber-coupled spectrograph and a two-element image slicer, in each echelle order the target spectrum consists of two closely spaced bright stripes. Below the target spectrum is a spectrum of the background with the same topology. However, as the background is much fainter than the target, it is visible only at wavelengths at which atmospheric or circumstellar emission lines are present. Except for a smooth continuum, the only conspicuous feature of the target spectrum itself is a broad emission line in the upper left quarter of the image. Image courtesy Otmar Stahl.

double subtraction scheme, no separate bias subtraction (or sky subtraction) is required.

6.4.2 Removal of Cosmics and Other Artifacts

All photon detectors are also sensitive to energetic charged particles. At observatory sites, such particles are present owing to natural radioactivity and cosmic rays. The contribution from radioactivity can be alleviated by avoiding the use of radiating materials close to the detectors. This applies, for instance, to optical glasses, which are often mildly radioactive. However, even if a spectrometer contains no radioactive materials, at observatory sites some energetic particle radiation is always present, and the resulting particle traces are recorded regularly by all photon detectors. In astronomy, the traces, which are caused by natural radioactivity and cosmic rays, are generally called *cosmics*, regardless of their true origin. Examples are shown in Figure 6.8. As shown in this figure, on images the cosmics appear as small bright spots. The individual traces normally consist of just one or only very few pixels. However, the affected pixels are often saturated. If the trajectory of a particle is close to the plane of the detector, a short streak is produced (as near the lower right corner of Figure 6.8). Even in this case, however, the profile of the cosmic perpendicular to its elongation is sharper than that of unresolved spectral lines, such as the narrow airglow line labeled A in Figure 6.8.

For pixels that are affected by comics, all target or background flux information is lost. However, because cosmics are (at least in one dimension) always sharper than the optical point-spread function, they can be identified reliably at practically any stage of the reduction process. Sometimes they are already detected and eliminated by an interpolation at the beginning of the reduction. However, interpolations can introduce systematic errors. Therefore, they are preferably removed at a later stage, at which (as described later) they can be eliminated safely without possibly introducing systematic flux errors.

In addition to cosmics, frames obtained with array detectors contain artifacts, which result from defective pixels or which have been caused by a previous overexposure. If the "bad pixels" have a lower sensitivity only, their effects are corrected by the flat-fielding procedure. The effects of dead and bright pixels can be removed together with the cosmics. Problematic are persistent charges from earlier (over-) exposures. Such persistent charges decay with time. However, because of this time variability, they are not properly corrected by normal flat-fielding. Charge persistence can occur in any array detector, but this phenomenon seems to be particularly widespread in InSb IR detectors (see, e.g., Campbell and Thompson, 2006).

6.4.3 Flat-Field Correction

Detector materials are never completely homogenous. Therefore, all imaging detectors show differences in their sensitivity between the individual pixels. Sometimes additional sensitivity variations are produced by reflection and interference effects at the various layers of solid-state detectors. Whereas the large-scale sensitivity variations normally are corrected using flux standard stars, the small-scale pixel-to-pixel variations are eliminated using spectra of a white screen that is illuminated uniformly (as far as possible) with a smooth-spectrum light source. Normally an incandescent lamp of an appropriate temperature is used for this purpose. The resulting spectra are called *flat fields*, or simply *flats*. To give good results, the light paths through the spectrometer of the flat-field exposures and the object spectra must be identical. Therefore, often flat-field screens in front of the telescope (attached to the dome structure) are used. However, experience shows that smaller screens inserted between the secondary mirror and the spectrometer can give results that are as good. Because the pixel sensitivity usually is a function of the wavelength, the flat-field light source should have the same color as the observed targets. Although this can be a problem for direct imaging, in spectroscopy the correct color of the flats can be ensured by taking the flats with exactly the same spectrograph setup as the target spectra. In this case, for both spectra the relations between the wavelength and the position on the detector are exactly the same. However, a color mismatch can occur if a significant amount of undispersed stray light reaches the detector. Because this can result in noise and large systematic errors, stray light suppression is one of the important design goals of modern spectrometers (see Section 4.4.6).

To derive directly the small-scale variations of the detector sensitivity, first "smooth flats" are generated by median-filtering the observed flats over an adequate range of pixels. Dividing the original flats by the smooth flats results in normalized flats that represent the (small-scale) relative pixel-to-pixel sensitivity variations. Corrected target spectra then can be obtained by calculating the quotient of the original target spectrum and the normalized flat.

For slit spectrometers, normally the whole detector surface is flat-field corrected. In the case of stationary fiber-coupled spectrometers, the flat-fielding can be restricted to the detector areas that receive light from the fibers, whereas "dark" regions on the detector are disregarded.

As in the case of the bias, in order not to increase the total noise, the normalized flats should have a significantly higher S/N than the target spectra. Normally the difference should be about a factor of ten. Thus, the flats must be well exposed and, if necessary, individual flats (of the same spectrograph setup) must be stacked until an adequate S/N is reached. If the spectrograph setup of the target observations can be reproduced sufficiently well, flat-field exposures again can be carried out in daytime following an observing night.

As pointed out in Section 4.6.1, the flat-fielding, as described earlier, also can correct flux variations caused by small-scale slit width fluctuations in laser-cut slit masks. However, an efficient correction of this effect requires that during the flat-field exposure the slits are projected to exactly the same position on the detector as during the object exposure. To meet this condition, flexure effects must be negligible and the robotic mask insertion into the focal plane must be sufficiently precise.

Flats can also be used to correct for the apparent flux variation with wavelength that is caused by the blaze function of a grating. An example for this

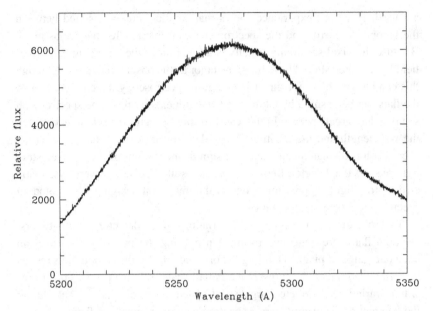

Figure 6.9. Raw spectrum of a star with an (in that wavelength range) essentially flat energy distribution. The apparent strong variation of the flux is due to the blaze function of the spectrograph grating.

application is given in Figures 6.9 and 6.10. As shown by these figures, in a well-behaved spectrometer the blaze function can be removed rather efficiently by dividing the spectrum by a flat-field frame. Of course, this procedure normalizes the spectrum only to that of the flat-field light source. To get a true flux scale, the flux calibration described in Section 6.4.8 must be applied.

6.4.4 Wavelength Calibration

Following the flat-field correction, a correct wavelength scale of the target and the background spectra must be established. For this purpose the wavelength calibration lamp spectra (see Section 6.3.1) are evaluated. First, using a suitable algorithm, the coordinates of the observed spectral lines of the calibration source on the detector are determined. In a next step the expected coordinates of these lines on the detector are calculated from the parameters of the spectrograph setup and the known wavelengths. In the case of MOS, this must be done separately for each individual slit (or fiber) position. Each observed line is then identified by searching in the observed spectra around the predicted coordinates

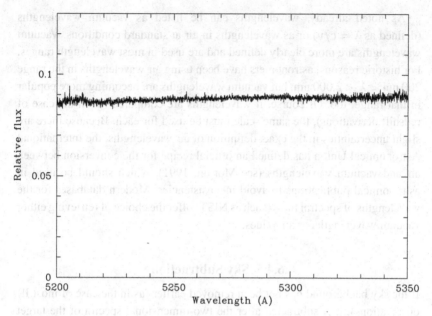

Figure 6.10. The spectrum of Figure 6.9 after division by a smooth flat-field frame. As shown in this figure, the division restores the original flat energy distribution and permits a much better assessment of the nature of the stellar spectrum. Spectral data courtesy Otmar Stahl.

until the lines are found. As the next step, a *dispersion relation* is established by fitting observed coordinates and wavelengths with a low-order polynomial. To exclude misidentifications, it is sometimes advisable to identify initially only the strongest lines to get a first-order wavelength scale. This initial scale can then be improved iteratively by including all lines, until a satisfactory accuracy is reached. In a last step, the dispersion relation derived for the comparison spectra is applied to the target spectra (including the sky background) and the target spectra are rebinned to a linear wavelength scale.

Depending on the initial sampling, the rebinning to a wavelength scale can reduce the spectral resolution and can result in an increased noise. Therefore, some authors prefer to skip the rebinning and to work with non-equidistant spectral tables. However, if suitable rebinning algorithms are used, adverse effects can be minimized and spectra with linear wavelength scales are easier to work with. The choice of the rebinning algorithm depends on what is offered by the particular image processing system that is used. Hints to the optimal rebinning can be found in the descriptions of these systems.

As noted already, wavelengths can be listed as vacuum wavelengths (defined as $\lambda = c/\nu$) or as wavelengths in air at standard conditions. Vacuum wavelengths are more clearly defined and are used in most wavelength ranges. For historic reasons, astronomers have been using air wavelengths in the range 200 nm $< \lambda <$ 2,000 nm, but vacuum wavelengths are becoming more popular in that range, too. Of course, if wavelengths are compared (as in the case of redshift derivations), the same scale must be used for each. Because there are slight uncertainties in the exact definition of air wavelengths, the International Astronomical Union has defined an official recipe for the conversion between air and vacuum wavelengths (see Morton, 1991), which should be used in astronomical publications to avoid inconsistencies. Modern databases for the wavelengths of spectral lines (such as NIST) offer the choice of retrieving either vacuum wavelengths or air values.

6.4.5 Sky Subtraction

If the sky background has not been removed earlier (as in the case of most IR observations), it is subtracted after the two-dimensional spectra of the target and of the background have been rebinned to a common wavelength scale. For each wavelength bin, the background can be determined from the sky pixels and subtracted from the corresponding pixels of the target spectrum. In the case of fiber-coupled instruments, this procedure is relatively simple, as the pixels illuminated by the target fibers and by the sky fibers are well defined. They can be determined accurately using the high-S/N flat-field frames. In the case of slit spectrometers, deriving the positions of suitable sky pixels is more complex, because the extent of the target and the positions of possible additional objects in the slit(s) must be determined before regions of undisturbed sky background can be identified. Faint wings of the target light distribution and faint background objects often are not directly detectable in the spectra. If deep direct images of the field exist, and if the absolute slit positions are sufficiently well known, these direct images can be used to identify pixels that are free of other objects and that represent the true sky background. Alternatively, all pixels along a slit can be added up in dispersion direction to obtain a deep *pseudo-white* image of the light distribution along the slit. (For obvious reasons, bad pixels must be excluded in these sums.) From the resulting mean "white" intensity distribution along the slit, it is usually possible to find regions of undisturbed background flux. If pixels that represent the undisturbed sky have been identified, they can be used to subtract the background from the object frames.

Disregarding the pixels that are affected by cosmics or detector defects, the subtraction of the background results in two-dimensional spectra that correspond to the true flux level in relative units. Because the sky has been subtracted, the signal level in these frames is lower than in the uncorrected frames. However, the noise level is not reduced by the subtraction. In fact, because any arithmetic operation of image frames tends to add noise, it may be increased by the sky subtraction. To keep the additional noise as small as possible, in each wavelength bin the background must be based on as many pixels as possible, and averages over these pixels must be used for the correction. Because the S/N increases with the number of averaged pixels (or the sky area involved), a longer slit results in a less noisy mean background and in a more accurate background-subtracted spectrum.

For faint objects, the total flux is always dominated by the sky background and the noise will be practically independent of the position in the slit. If a noise increase can be avoided, this noise level will not be changed by the subtraction, but the S/N will always be reduced.

6.4.6 Extraction of One-Dimensional Spectra

After the background has been subtracted, in principle, one-dimensional spectra can be extracted by adding up the flux (in ADUs) of all pixels of a given wavelength bin. Of course, pixels that contain no target flux can be omitted, and pixels containing flux from background objects have to be excluded in these sums. With these precautions, the procedure gives the total target spectral flux (in arbitrary units) as a function of the wavelength, provided that there are no cosmics or other artifacts that falsify the flux locally.

Although the simple addition gives (under the conditions mentioned previously) a systematically correct result, it tends to produce one-dimensional spectra with an unnecessary high noise. This is due to the fact that in a simple addition, the pixels with the highest object flux and best S/N are added with the same weight as the pixels at the edge or outside of the object area, which contribute mainly (or only) noise. Therefore, to retain the S/N of the data, the pixels must be added up taking into account their individual S/Ns. The optimal way to accomplish this objective depends on the spatial extent of the spectra and on the instrument. Various different procedures have been proposed in the literature. Most frequently used is an algorithm described by K. Horne (1986) for the extraction of spectra of point sources recorded with CCD detectors. The Horne algorithm can be adopted to small extended objects, and it may also be applied to other types of imaging detectors. Because of its importance in astronomy, the Horne method will be briefly summarized here. A comprehensive description of the details can be found in Horne's original 1986 paper.

The Horne algorithm uses as input the wavelength-calibrated two-dimensional spectra. It adds up the pixels of each wavelength bin with weights that are proportional to the square of the pixel flux divided by the variance of the pixel values. For this purpose, smooth functions are derived that describe the variance of the pixel values and the profile of the flux distribution perpendicular to the dispersion. The smooth variance functions are obtained by averaging the variance over a suitably large area around a given pixel. To obtain smooth functions representing the flux distribution perpendicular to the dispersion, the variation of the normalized flux profile along pixel rows *in dispersion direction* is approximated by low-order polynomials. These polynomials are positive for pixels where flux is observed, and zero elsewhere. Because many data points contribute to these smooth functions, they are much more accurate than the individual pixel values. Thus, these smooth functions do not significantly affect the noise of the final spectra.

Cosmics and other defects are eliminated by removing all pixels that deviate by more than a predetermined number of standard deviations from the flux profile functions. After removing a pixel, the corresponding profile function is recalculated without the removed pixel. This procedure is called σ *clipping*. Because the mean variance becomes smaller after the removal of a bad pixel, σ clipping may have to be repeated. Horne suggests a removal of pixels that deviate by more than $4\,\sigma$. However, depending on the detector, different factors can give better results. Pixels that are removed by the clipping get zero weight in the addition. Because for each wavelength bin the total weight is normalized to 1, neglecting cosmics and bad pixels has no systematic effect on the total flux, but the S/N is reduced at the corresponding wavelength.

The variance functions and total weights also provide information on the accuracy of the extracted spectra. These data can be used in the error estimates of the final results by following the error propagation through the reduction process. However, this error propagation is complex and errors derived by comparing multiple frames normally give more reliable results. A thorough discussion of this issue has been given by Noll (2002).

The Horne algorithm works well in its original form if the profile of a spectrum perpendicular to dispersion varies smoothly with wavelength. For other targets (such as extended emission line sources, where the extent differs for the individual lines) different procedures may be needed, but the basic principles of the Horne algorithm are still applicable.

In fiber-coupled stationary spectrographs, the flux profiles perpendicular to dispersion normally are very stable and can be determined accurately from the high-S/N flat-field exposures.

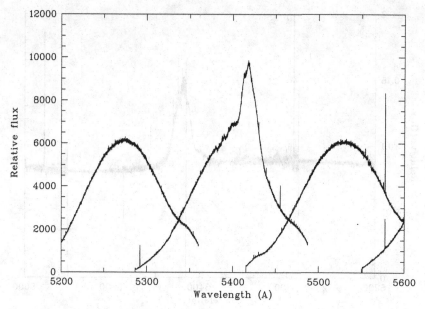

Figure 6.11. Three adjacent echelle orders uncorrected for the grating blaze function.

6.4.7 Merging of Echelle Orders

The reduction of echelle spectra follow the same basic steps as those described for other spectra taken with array detectors. The wavelength calibration of echelle frames can be carried out for the whole frame or separately for each order. Because many more lines are involved, a dispersion relation based on a total frame normally is more accurate. The extraction of the one-dimensional spectrum has to be done for each order separately. Then the resulting individual orders have to be merged into a single spectrum.

Before the orders can be merged, they have to be corrected for their blaze functions. As described in Section 6.4.3, this can be done using the smooth (not normalized) flats. As shown by a comparison of Figures 6.11 and 6.12, if the flats are free of systematic errors, the flat-fielded orders agree well in their overlap regions. However, because of the low flux in the extreme wings of the blaze function, the contributions from these regions may have a very low S/N in the merged spectrum, and uncorrected bad pixels in the wings are strongly amplified. These extreme wings do not provide a significant contribution to the signal, but they can cause systematic errors due to incompletely corrected

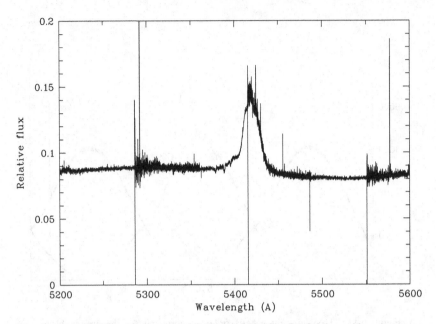

Figure 6.12. The echelle orders of Figure 6.11 after flat-fielding with a (not-normalized) flat. Note the agreement of the different orders in the overlap regions.

bad pixels Therefore, they are normally omitted from the merged spectra. For the remaining pixels in the overlap regions an optimal result is obtained by calculating weighted averages, in which the local $(S/N)^2$ is used as a weight function. Figure 6.13 shows a corrected merged spectrum of the orders of Figure 6.11.

6.4.8 Flux Calibration

Depending on the specific reduction code, the extraction procedures results either in a relative energy or in a relative flux as a function of wavelength. If the output of the reduction program corresponds to an energy, the relative flux can be calculated by dividing the energy by the effective integration time.

To convert the relative flux spectra $F_R(\lambda)$ of the target into physical flux units, the spectra of the flux standard stars (see Section 6.3.2) must be reduced in exactly the same way as the target spectra. This results in standard-star spectra expressed in relative flux units $F_{SR}(\lambda)$. In the case of observations from space, or if a target and the standard star have been observed in a photometric night at the same zenith distance, and if both spectra have not suffered flux losses at

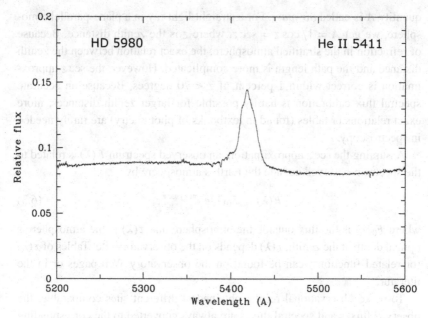

Figure 6.13. The echelle orders of Figure 6.11 after flat-fielding, merging, and removal of bad and low-S/N pixels and of high-frequency noise. Shown here is the profile of the He II 5411 Å line in the spectrum of the Wolf-Rayet+OB binary HD 5980 (see, e.g., Koenigsberger et al., 2011) in the Small Magellanic Cloud. Spectral data courtesy Otmar Stahl.

the slit, a flux-calibrated target spectrum $F(\lambda)$ can be calculated according to

$$F(\lambda) = \frac{F_R(\lambda)}{F_{SR}(\lambda)} F_S(\lambda), \tag{6.2}$$

where $F_S(\lambda)$ is the spectrum of the flux standard in physical units.

Flux standards normally are observed with a wide slit and, therefore, can be assumed to be free from slit losses. However, for most targets, narrower slits are used to get a higher spectral resolution and a better S/N. Thus, in the case of point sources, the wings of the seeing pattern are often truncated. At the major observatories, the seeing is monitored during each clear night. These monitoring data can be used to model the slit losses and correct the observed flux accordingly. For extended sources, slit loss corrections are more complex and require some knowledge about the light distribution of the object.

To correct for different zenith distances between the target and a flux standard, the different effective path length l_z through the atmosphere must be calculated. In astronomy, this effective path length normally is expressed by the ratio $A = l_z/l_{z=0}$, where $l_{z=0}$ is the path length in the zenith. The dimensionless quantity A is called *air mass*. For a straight light ray in a plane-parallel atmosphere, we get $A = 1/\cos z = \sec z$, where z is the zenith distance. Because of refraction in the stratified atmosphere, the exact relation between the zenith distance and the path length is more complicated. However, the $\sec z$ approximation is correct within 1 percent, if $z < 70$ degrees. Because an accurate spectral flux calibration is hardly possible for larger zenith distances, more exact relations or tables (found in textbooks of photometry) are rarely needed in spectroscopy.

Assuming the $\sec z$ approximation, an observed spectrum $F(\lambda)$ is related to the spectrum observed outside the Earth's atmosphere by

$$F(\lambda) = F_0(\lambda)e^{-\tau(\lambda)\sec z}, \qquad (6.3)$$

where $F_0(\lambda)$ is the flux outside the atmosphere and $\tau(\lambda)$ is the atmosphere's optical depth at the zenith. $\tau(\lambda)$ depends on the observatory site. Tables of $\tau(\lambda)$ (or related functions) can be found on the observatory Web pages or in the literature.

To make observational results obtained at different sites comparable, the observed fluxes and spectral fluxes are always converted to the corresponding values outside the atmosphere. By inverting Equation 6.3, we get for this conversion

$$F_0(\lambda) = F(\lambda)e^{\tau(\lambda)\sec z}, \qquad (6.4)$$

where z is the effective zenith distance of the observation. Because the integration times of spectral observations may be long, and because the atmospheric extinction varies nonlinearly with z, a simple average of the zenith distance at the beginning and the end of the integration often does not give adequate results. A frequently used better approximation to the effective air mass is

$$A_{eff} = (A_b + 4A_m + A_e)/6, \qquad (6.5)$$

where A_b, A_m, and A_e are the air mass values at the beginning, halfway through, and at the end of the exposure, respectively (for a derivation see Stetson, 1989).

The accuracy of a flux calibration depends on the quality of the atmospheric extinction correction, on the slit-loss correction, and on how well the details of the standard star observations match the target observations. Photometric observations obtained from sky images in standard filter bands normally provide more accurate values of the integrated flux in the corresponding filter band. A numerical integration of flux-calibrated spectra over the same wavelength band obviously should give the same result. If the results do not agree, usually the photometric results are assumed to be the correct ones. Because the difference normally is an (essentially) wavelength-independent offset, spectra are often corrected by normalizing the results to such photometric value. Today, photometric data for a large number of stars can be found in the astronomical data collections. Therefore, it is usually not necessary to obtain new photometric measurements for this type of correction.

6.4.9 Corrections for Atmospheric Absorption Features

In Section 6.4.5, the procedure used to correct observed spectra for the sky background contribution was described. This correction removes the emission features produced in space and by the Earth's atmosphere, including interplanetary and atmospheric stray light and the emission of atmospheric atoms and molecules. However, the sky subtraction does not remove absorption features that are produced by the Earth's atmosphere.

The main constituents of our atmosphere are nitrogen and oxygen molecules. In the visual spectral range, there are only few absorption features of O_2. However, many O_2 and N_2 lines are present in the UV and IR bands. Because the atmospheric oxygen and nitrogen fraction is essentially constant, for a given site these lines show few variations. Their strength tends to be proportional to the air mass, and their profiles are well known. Therefore, these lines normally can be removed from target spectra using numerical modeling.

In the IR spectral range there are many molecular absorption features due to H_2O, CO_2, CH_4, and other common atmospheric molecules. Because the concentration of these molecules varies with time, the strength of their lines must be determined empirically close to the target observations. In principle, flux standard star observations can be used for this purpose. However, the limited number of bright standard stars often does not permit carrying out a sufficiently number of such observations during an observing night. Therefore, during IR spectroscopy runs, additional short observations of bright stars are often carried out to determine the atmospheric absorption lines. The stars used for this purpose (sometimes called *telluric stars*) need not to be flux standards, but their spectra must be known on a relative scale. Because the solar spectrum is well known, stars of the solar spectral type can be used. After correction for their intrinsic spectra, the observed telluric-star spectra give directly the strength of the atmospheric absorption lines at the corresponding time and zenith distance.

6.4.10 Reference Frame Corrections

The wavelength scale that is derived by means of the spectral lines of a calibration lamp corresponds to the spectrograph's reference frame at the time of the observations. To make spectroscopic results comparable, these wavelengths must be converted to a standard reference frame. Several different standard reference frames are in use in astronomy. Most frequently, the wavelengths of stellar spectra are given for the barycenter of the solar system (called the *barycentric standard of rest*, or BSR) or the center of the Sun. Values given for these systems are called *barycentric* and *heliocentric*, respectively. Because the Sun moves slowly relative to the solar-system barycenter, the difference between these two systems is small and the distinction is often neglected in the literature.

To convert observed wavelengths and radial velocities to the barycentric system, the Doppler effects due to the Earth's rotation, the Earth's orbit, and the gravitational potential difference must be taken into account. Although the computation of the correction for the Earth's rotation is simple, calculating the effect of the Earth's orbital motion is a bit more complex. For its calculation the algorithm proposed by P. Stumpff (1980) is usually used. Corresponding subroutines can be found in the astronomical data processing systems mentioned above. In IRAF, the package RVSAO can be used for this purpose.

In galactic astronomy, the mean motion of the galactic matter in the solar neighborhood is often used as a reference frame. This frame is called the *local standard of rest* (LSR). Because the Sun has a peculiar motion (of 13.4 km/s), LSR and BSR are not identical.

Radial velocities of extragalactic objects often are converted to values that would be observed from the center of our galaxy.

6.5 Archiving Spectral Data

Spectra can contain a vast amount of information. Often, the observer is interested only in specific aspects of this information, and only this fraction of the information will then be extracted and published. The remaining information content, however, may well be of great value for other astronomers. Moreover, in the case of variable objects, spectra taken a long time ago sometimes result in important scientific breakthroughs decades after the observations took place. Therefore, practically since the beginning of photographic spectroscopy, astronomers have been archiving their spectra. Initially, the spectral archives simply consisted of collections of spectral plates. With the development of photoelectric detectors, digital storage devices became the main medium of spectral archives. Experience shows that – with some precautions – developed and (well) fixed photographic plates can be stored without significant deterioration for at least one century. The effective lifespan of digital storage devices is less well known. However, at least some of the electronic storage media tend to become unreadable on a much shorter timescale. Apart from the durability of the media, digitally stored data also suffer from the rapid progress of the IT hardware and software, which often makes it impossible to read old media or data files with present-day computers. According to various estimates, since the introduction of digital storage devices more astronomical data have been lost than during the whole photographic era.

At the time being, the loss of information due to deteriorating storage media can be solved only by recopying the data regularly to new (and, hopefully, longer-lasting) devices. This is being done regularly at various large observatories and astronomical data centers. However, ensuring the readability of data also requires that storage formats are used that are independent of a specific hardware and of a specific operating system. In astronomy, such a commonly agreed data format exists since 1981 in the form of the Flexible Image Transport System (FITS). This system is endorsed by the International Astronomical Union and is maintained by the NASA FITS support office (fits.gsfc.nasa.gov), which provides detailed information on the current version of FITS. Today, FITS is used almost universally in astronomy, and interfaces exist to all important astronomical data software packages. Moreover, FITS is becoming popular outside astronomy, and several general-purpose image processing systems can process FITS data as well. All major observatories distribute spectroscopic results as FITS files. Although FITS is being continually improved, care is taken that files based on earlier versions can still be processed.

Each FITS file consists of two parts. The second part contains the actual data in the form of a table or a digital image. This information is given in a binary format. The first part (the "header") contains metadata, which provide vital explanatory information on the table or image part. Among the header information are the name of the observed object, the type of observation, the date, the history of the data, and so on. The first part is coded in ASCII format. Thus, it can be read directly without format conversion. In detail, the header contains a series of lines consisting of a keyword and the corresponding parameters. (For details, see the FITS Web page cited previously.) It is obvious that any new astronomical spectrometer should be designed to deliver FITS files as the final data output.

Extensive archives of spectral data exist at all large observatories and at astronomical data centers, such as the Centre de Donée Astronomiques de Strasbourg (CDS, cdsweb.u-strasbg.fr) in France. An important depository of infrared data is the Infrared Processing and Analysis Center (IPAC, www.ipac.caltech.edu) of NASA. NASA also operates large archives for HST spectra (www.stsci.edu) and other data obtained from space facilities.

To spare the astronomers a search in many different data centers, at present a system of virtual observatories is under development, which aims at providing a uniform and convenient access to all existing digital (or digitized) astronomical data, including many spectra. Apart from retrieving spectra of interest, observers also can archive their spectral results in these virtual observatories (as well as in data centers, such as the CDS). Information on the present status of the virtual observatories and links to the individual members of this cooperation can be found on the Web page of International Virtual Observatory Alliance (IVOA, www.ivoa.net).

As a result of the developments described in the preceding paragraphs, today a large amount of spectral information can be collected without using a telescope. If the access to the archive data becomes as convenient as planned by the organizers of the virtual observatories, archives will play an important role for the future of astronomical spectroscopy.

7

UV, X-Ray, and Gamma Spectroscopy

This chapter covers astronomical spectroscopy at wavelengths between the UV and the gamma rays, which corresponds to a range of photon energies between about 4 eV and about 10^{14} eV. Naturally, different techniques are used at the extreme ends of this vast frequency range. However, the methods change continuously with the photon energy, and there exist extended regions in which the different methods overlap. Moreover, some of the basic problems and solutions are common throughout this range. In the literature, the high photon-energy range is often divided into the *near ultraviolet* (NUV, wavelengths about 200–380 nm), *far ultraviolet* (FUV, 100–200 nm), *extreme ultraviolet* (EUV, 10–100 nm), *soft X-rays* (0.5–10 nm), *hard X-rays* (2.5 pm–0.5 nm), and *gamma rays* (<2.5 pm). For convenience, these subdivisions will also be used in this text.

Energetic photons can interact with matter in many different ways. Among the relevant physical processes are the ionization of matter, the photoeffect, Compton scattering, and (at energies > 1 MeV) electron–positron pair production. All these processes can absorb over broad continuum bands. Because of this high absorption probability of energetic photons, optical techniques, which are based on the refraction and reflection of light, either cannot be used at all or require special layouts. The effective absorption cross section decreases again for the very high photon energies of hard X-rays and gamma rays. However, at these high energies, the radiation penetrates the normally reflecting materials, and the refractive index of all materials is uniformly very close to 1. Therefore, at these wavelengths optical methods are also excluded.

In spite of these fundamental difficulties, during the past decades special optical methods have been developed, which can be applied to UV and X-ray spectroscopy. The first section of this chapter provides an introduction to these special techniques. It is followed by a discussion of spectroscopic instrumentation that has been developed making use of this type of optics. In the final part of this chapter, methods are described that are employed for the spectroscopy of the hard X-rays and the gamma rays, for which optical methods (so far) are not applicable. (Readers interested in a more comprehensive treatise on EUV and X-ray spectroscopy can find additional information in the *Handbook of X-Ray Astronomy* by Arnaud et al. (2011).) A useful collection of basic physical data, which are relevant for high-energy astronomy, can be found in the *X-Ray Data Booklet*. This very informative booklet has been compiled by colleagues of the Lawrence Berkeley National Laboratory (LBL). It can be downloaded freely from the LBL home page (http://xdb.lbl.gov/xdb.pdf).

NUV observation at wavelengths >320 nm can be carried out from the ground. For shorter wavelengths, the Earth's atmosphere is essentially opaque owing to molecular and atomic absorption bands. Thus, observations shorter than 320 nm must be made from space. An exception is the gamma range above about 0.1 TeV, at which the interaction of gamma photons with atomic nuclei in the high atmosphere can be used to detect and to measure this hard radiation.

An astrophysically interesting, but particularly difficult, range is the EUV region. Because of very high absorption cross section of atomic hydrogen short of the Lyman series limit (at 91.2 nm), much of the galactic interstellar space is opaque at these wavelengths. As a result, in the EUV, only nearby objects and selected lines of sight are accessible to astronomical observations.

7.1 UV and X-Ray Optics

As noted previously, UV and X-ray photons are strongly absorbed in most materials, and for X-rays the refractive index becomes $n \approx 1$. This makes refractive optics difficult or impractical. Therefore, dioptric optical systems are rarely used for wavelengths <300 nm. Even in the NUV range (300–380 nm), special UV-transparent glasses or crystals must be employed, which are expensive, delicate, and difficult to polish. As a result, reflecting optical components are preferred throughout the UV range, and they are the only choice for X-ray astronomy.

Reflective optics is being used for photon energies up to about 80 keV (i.e., for wavelengths >16 pm). This range may be extended in the future. At NUV and FUV wavelengths, normal-incidence mirrors can be employed, whereas spectroscopy at shorter wavelengths requires grazing-incidence optics.

7.1.1 Normal-Incidence Mirrors

In the visual spectral range, the standard coating material for telescope mirrors is aluminum. Pure aluminum has a reflectivity of >85 percent over a very broad wavelength range, extending well into the UV. At normal incidence, the reflectivity of a freshly evaporated pure aluminum film extends to almost 100 nm. However, in the presence of oxygen, the surface immediately forms a thin (1–5 nm) aluminum oxide film, which is transparent in the visual but absorbs photons with wavelengths <300 nm.

Because FUV observations must be done from space anyway, one might consider bringing a bare aluminum mirror into space by keeping it at vacuum conditions until in orbit, or by coating the mirror in space. However, apart from the technical difficulties involved, even a small amount of oxygen from the outgassing of the spacecraft or from the rest atmosphere in a low Earth orbit could rapidly destroy the UV reflectivity of such a mirror.

A more practical solution is protecting new aluminum coatings by a thin overcoating of a UV-transparent material. Suitable protective coatings are magnesium fluoride and lithium fluoride. MgF_2 coatings can be used for wavelengths above about 120 nm, and Al/LiF coatings extend practically to the Al cutoff near 100 nm.

At EUV wavelengths, where protected aluminum becomes inefficient, high-atomic-weight metals, such as iridium or osmium, are used as mirror coatings. However, the reflectivity of these materials is much lower than that of aluminum at longer wavelengths. The reflectivity of iridium starts at $\lambda = 30$ nm, and at wavelengths ≥ 40 nm iridium has a reflectance of about 20 percent. For EUV wavelengths $\lambda > 50$ nm, silicon carbide (SiC) is frequently used. For wavelengths ≥ 50 nm, the reflectivity of SiC is about 40 percent with a relatively flat wavelength dependence, but at shorter wavelengths the reflectivity of SiC drops to practically zero. At present, no materials are known that provide a broadband normal-incidence reflection at wavelengths <30 nm. For all materials, the reflectivity depends on the angle of incidence. Usually higher reflectivities can be reached with oblique reflections.

To work properly, reflecting optical components must have a roughness and a deviation from the ideal surface that is small relative to the wavelength of the incident light. Because X-ray wavelengths correspond to the size of atoms, it is difficult or impossible to meet this condition for hard X-rays. As a result, producing (and testing) reflecting optical surfaces for UV and X-ray radiation is more demanding than in the visual or the IR, and the manufacturing tolerances of the surfaces often limit the reflectivity. As a consequence, the achievable values of the reflectivity depend not only on the coating material, but also on the coating process and on the smoothness of the substrate. Exact reflectivity values must be obtained for each individual component, and values given in the literature can be guidelines only.

With the coating materials discussed previously, reflecting optical components of UV spectrographs, such as collimators, cameras, and reflection

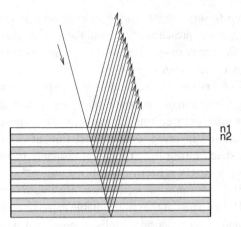

Figure 7.1. The principle of a multilayer mirror for EUV and X-ray optical systems.

gratings, can be constructed following the principles outlined in Chapter 4. Naturally, because of the low reflectivities in the UV, unnecessary reflections must be avoided. Examples of such conventional UV spectrometers are described in Section 7.2.

7.1.2 Multilayer Mirrors

In laboratory application of X-rays (such as X-ray microscopy), reflecting or refracting optical elements are often replaced by devices that are based on diffraction effects. Examples are Fresnel zone plates and Fresnel lenses, which work well even at high X-ray energies. However, Fresnel optics gives sharp images only for monochromatic radiation. In the laboratory it is possible to generate intense quasi-monochromatic X-ray beams, but astronomical sources normally have broad continuous spectra, and they are always faint. There exist astronomical X-ray sources that emit mainly emission lines, but these sources tend to have multiple lines with different wavelengths. Therefore, in X-ray astronomy, apart from gratings, diffractive optical components are rarely used.

An exception are *multilayer mirrors*, which have been employed successfully in several solar space observatories, and which play an important role in the context of the grazing-incidence optics, which is discussed in the next subsection.

The structure and the functioning of multilayer mirrors is outlined in Figure 7.1. A mirror of this type consists of a large number (typically several

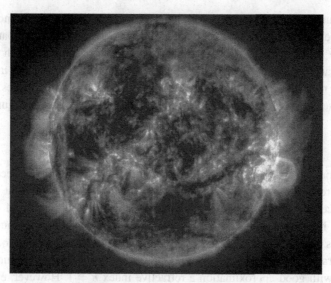

Figure 7.2. Example of narrow-band ($\lambda/\Delta\lambda \approx 10^2$) EUV image obtained with a normal-incidence multilayer reflecting telescope. Shown is an image of the Sun in the 17.1 nm Fe IX/X emission line. Image courtesy SOHO/EIT consortium (ESA and NASA).

hundreds) of alternating plane-parallel layers of transparent materials with different refractive indices $n1$ and $n2$. To make the difference $n1 - n2$ as large as possible, layers of high and low atomic-weight materials are combined. Typical heavy-weight materials are molybdenum, tungsten, or platinum, whereas Si or SiC are used as low-weight species. Because a discontinuity of the refractive index results in a partial reflection, each incident light ray results in a large number of parallel reflected rays. By choosing the thickness of the individual layers accordingly, an optical path difference between adjacent rays of just one wavelength can be produced. For the wavelength that meets this condition we get constructive interference of all reflected rays. Therefore, the device works as an efficient mirror for this wavelength.

As noted already, at X-ray wavelengths the refractive index of all materials tends to be close to unity. Thus, the difference $n1 - n2$ is always small. The intensity of the individual reflected rays of a multilayer mirror is therefore very low. However, because the individual layers are thin (of the order $\lambda/2$), losses are moderate and many reflections are possible before the beam is absorbed. Therefore, mirrors consisting of a few hundred layers can reach reflection efficiencies as high as 70 percent.

Because of the wavelength dependence of constructive interference, normal-incidence multilayer mirrors can be used only in narrow wavelength bands of typically $\lambda/\Delta\lambda \approx 100$. Moreover, a different mirror is needed for each wavelength band. Because of this restriction, normal-incidence multilayer mirrors are used mainly to obtain quasi-monochromatic EUV and X-ray images in the light of selected emission lines. A notable example is reproduced in Figure 7.2.

7.1.3 Grazing-Incidence Optics

In transparent materials, at most wavelengths the refractive index decreases with increasing wavelength. Exceptions are found near wavelengths at which resonances between the light waves and the electrons in a material occur. At very high frequencies, resonance effects become negligible because the electrons cannot react to very fast oscillations. Therefore, the phase velocity of gamma rays in matter does not deviate significantly from that in a vacuum, and we get with good approximation a refractive index $n = 1$. However, even at these high energies the refractive index decreases with increasing wavelength. Therefore, for X-rays we get $n < 1$. According to Equation 3.3, this means that the phase velocity of X-rays in matter c_m exceeds the velocity of light in vacuum c. (This is not a contradiction to special relativity, as the group velocity of the waves and the velocity of photons do not exceed c.) As derived in textbooks of electrodynamics, the deviation from $n = 1$ follows approximately a relation $1 - n \propto \lambda^2$. Typical values for the refractive index at a photon energy of 30 keV are $1 - n \leq 10^{-6}$. As noted earlier, this very small deviation from $n = 1$ makes refractive optics impractical (at least for astronomical applications). On the other hand, the refractive index $n < 1$ can be used to design reflective optical components for wavelengths at which, owing to surface absorption, normal-incidence mirrors are no longer feasible. These optical components make use of the total reflection effect.

At visual wavelengths (internal) total reflection takes place when a light ray propagating in a medium with a refractive index $n > n_{air}$ encounters the surface with an inclination angle that, according to Snell's law (Equation 3.27), does not permit a refraction into air. In this case, the ray is reflected back into the medium practically without losses. For glass with a refractive index of $n = 1.5$, this occurs for inclination angles larger than 42 degrees. As noted in Chapter 4, at visual wavelength the total-reflection effect is used to confine light rays in optical fibers. Because, as explained earlier, X-ray instruments must be operated in the vacuum of space, the refractive index $n < 1$ can be used to prevent rays entering matter, where they would be absorbed. Moreover,

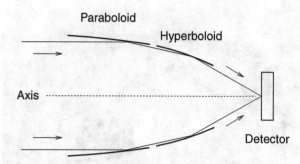

Figure 7.3. Schematic layout of a type I Wolter telescope.

in this way one can make use of the practically 100 percent efficiency of total reflection. On the other hand, because for X-rays we have $n \approx 1$, for all materials the critical inclination angle is close to 90 degrees. Thus, for high photon energies, the beam must be almost tangential to the reflecting surface to achieve total reflection. Optics based on this effect is called *grazing incidence* or *glancing incidence* optics. Although at X-rays we always have $n < 1$, the exact value of the refractive index and, consequently, the critical angle for the onset of total reflection vary greatly with the material.

The most important and best-known examples of grazing incidence optics are the Wolter X-ray telescopes, which were suggested by the German physicist Hans Wolter (1911–1978). A Wolter telescope always consists of two axis-symmetric surfaces formed by conic sections. The most common (Wolter type I) design is outlined in Figure 7.3. It consists of a paraboloid followed by a hyperboloid with the same rotational axis. Because all rays that are incident parallel to the axis of a paraboloid intersect at the paraboloid's focus (regardless of the position at which they meet the parabolic surface), in principle, a paraboloid alone can be used to produce an image. However, simple parabolic mirrors result in aberrations for beams that are not exactly parallel to the optical axis. These aberrations become very severe for the grazing-incidence geometry. Therefore, the Wolter type I telescope combines a parabolic and a hyperbolic mirror in the same way as the well-known Cassegrain systems, which are used at longer wavelengths. In contrast to a simple paraboloid, the Wolter telescope can image a sizable field around the focus point with an acceptable image quality.

Because only the rays are reflected that are almost (within about a degree) tangential to the reflecting surface, the effective entrance aperture of a Wolter telescope is a narrow annulus. The space inside this annulus is not used.

Figure 7.4. Parabolic component of one of the four Wolter type I telescopes of the Chandra X-ray observatory. The mirror is shown here before being coated with iridium. The blank consists of Zerodur and has a diameter of 123 cm. Image courtesy NASA/CXS/SAO.

Therefore, several (or many) Wolter telescopes of different diameters can be nested, and their output can be combined in a common focus, as indicated in Figure 7.7. Figure 7.4 shows the paraboloid of a large Wolter telescope. This paraboloid is the largest of eight grazing-incidence mirrors that were manufactured for the Chandra X-ray observatory.

Obviously, Wolter telescopes can be used as optical elements in the same way as bi-convex lenses and concave mirrors at longer wavelengths. The grazing-incidence principle can also be applied to other reflecting optical components, such as reflection gratings.

Grazing incidence optics is used in the EUV and for X-ray instruments. With heavy-metal coatings, such as iridium, grazing incidence optics can be constructed for photon energies up to about 15 keV. For higher energies, reflecting optical components can be realized by combining the grazing incidence concept with the multilayer coatings that were described in the previous subsection. An example of the use of multilayer coatings in Wolter telescopes is the NuStar mission of NASA (http://www.nustar.caltech.edu). With Pt/SiC and W/Si multilayers, NuStar is usable in the photon energy range between 5 keV and 80 keV. (At 80 keV the efficiency drops steeply owing to K-shell absorption of Pt.) Multilayers that may make it possible to apply grazing-incidence optics to even higher photon energies are under study.

7.2 UV Spectrometers

UV spectrometers for wavelengths >300 nm can be built with reflective optics using the same type of components and the same basic designs as those described for the visual and the IR ranges. These designs can be extended to most of the FUV range if reflective optics with protected aluminum is used. Most spectrometers for wavelengths >115 nm employ mirrors with Al/MgF_2 coatings, as MgF_2 is rather stable in vacuum.

As noted before, UV observations below 320 nm must be carried out from space. Because of the high costs of space experiments and the technical problems of EUV/FUV optics, only relatively few astronomical spectrometers have been operated in this range. Most of these instruments have been flown for a limited period and are no longer operational. Thus, astronomers interested in UV spectra often must rely on data that were obtained in the past with such instruments and are still accessible through the data archives of the agencies that launched and operated the historic UV missions.

As noted in Section 1.4.3, the most important instrument for the development of FUV spectroscopy was the International Ultraviolet Explorer (IUE) satellite. IUE was operated jointly by NASA and ESA between 1978 and 1996. It had a 45-cm protected-aluminum primary mirror, and it covered the wavelength range 115–330 nm using two echelle spectrometers with resolutions $R = \lambda/\Delta\lambda$ around 10^4. Moreover, by using the cross disperser alone, low-resolution spectra of $R \approx 350$ could be produced.

The most efficient instrument among the currently active spectrometers for the UV wavelengths ≥ 115 nm is the Cosmic Origins Spectrograph (COS) of the Hubble Space Telescope (HST; see www.stsci.edu/hst). Using a set of different reflection gratings, COS can reach spectral resolutions in the range $1,500 < R < 24,000$. Because of the larger (2.4-m) primary mirror of the HST and better detectors, COS can reach much fainter objects than had been possible with IUE and other past FUV missions. The HST spectrograph STIS has an FUV capability too, but the sensitivity of COS is much superior at these wavelengths.

Another currently active space facility with an FUV spectroscopic capability is the GALEX observatory. GALEX is mainly an imaging mission for the 135–300 nm range. However, by inserting a CaF_2 grism into the convergent beam, slitless low-resolution spectra can be obtained over the full field of the two GALEX cameras. The spectral resolutions are $R = 90$ for the long-wavelength (177–283 nm) band and $R = 200$ for the short-wavelength (134–179 nm) band of GALEX. Detailed information on this instrument can be found on the GALEX Web page (www.galex.caltech.edu).

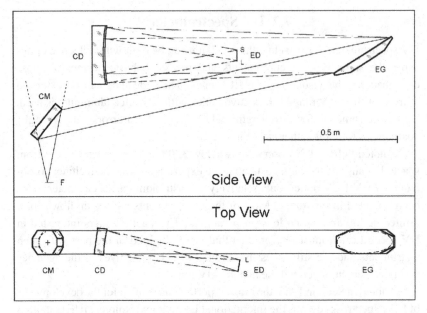

Figure 7.5. Optical layout of an FUV echelle spectrometer. To reduce the light losses at the reflecting surfaces, the collimator is an off-axis paraboloid illuminated at a 103-degree angle, and the cross disperser and the camera are combined in a single concave mirror. From Barnstedt et al. (1999).

Historically the most important instrument for the FUV/EUV range between 90 nm and 120 nm (corresponding roughly to the range of the Lyman series of atomic hydrogen) has been the Far Ultraviolet Spectroscopic Explorer (FUSE) satellite (http:// fuse.pha.jhu.edu). FUSE was launched in 1999 and took data until 2007. Its four polygonal 35-cm telescopes were feeding their light into four separate (slightly modified) Rowland spectrographs (see Section 4.4.5). Two of the telescope–spectrograph combinations were optimized for the 90–110 nm range, using SiC-coated mirrors, whereas the two other branches were optimized for 100–120 nm wavelengths by using Al/LiF mirrors. Because in each branch there were only two reflections (the main mirror and the Rowland grating), FUSE had a very good throughput (for these wavelengths). Moreover, the large gratings of FUSE (with a 1.62-m radius of curvature) resulted in spectral resolutions up to $R \approx 30,000$.

So far only very few astronomical spectrometers have been operated in the EUV range < 90 nm. The pioneering experiment at these wavelengths was the Extreme Ultraviolet Explorer (EUVE), which was launched by NASA in 1992 and which studied the 7–76 nm sky. For this purpose, EUVE employed

Figure 7.6. FUV (90–140 nm) echelle spectrum of the O9.5V star HD 93521 obtained with ORFEUS. At the left margin, order numbers are indicated. From Barnstedt et al. (1999).

four (Wolter type I or II) grazing-incidence telescopes and three spectrometers, giving a resolution $R \approx 260$. The mission resulted in an EUV spectral atlas that was published by Craig et al. (1997). (This publication also provides details on the design of the EUVE spectrometers and links to additional literature.)

Higher-resolution EUV spectra were obtained with the U.S./German ORFEUS experiment, which was launched for two short-term flights in 1993 and 1996 using the retrievable space platform ASTROSPAS (see Grewing et al., 1991). ORFEUS had an iridium-coated 1-m mirror and two spectrographs. Using a Rowland spectrometer (built by a group at University of California, Berkeley), spectra with $R \approx 3,000$ could be obtained in the wavelength range 40–115 nm. The second instrument was an FUV echelle spectrometer that produced spectra with $R \approx 10^4$ in the 90–140 nm range. Figure 7.5 shows the optical layout of the FUV echelle instrument. A high-resolution echelle spectrum obtained with ORFEUS is reproduced in Figure 7.6.

7.3 Photon-Energy Sensitive X-Ray Detectors

As described in Section 1.4.2, the first low-resolution spectra of cosmic X-ray sources were taken with proportional counters and similar gas ionization detectors. These devices became obsolete for spectroscopic applications after X-ray-sensitive Si-based CCD detectors were developed. X-ray CCDs soon became the most important photon-energy sensitive detectors for energies $h\nu \leq 15$ keV. Modern versions (with slightly different technologies) were included in the large X-ray observatories Chandra (operated by NASA) and XMM-Newton (operated by ESA). The detector systems of both these missions – named Chandra Advanced CCD Imaging Spectrometer (ACIS) and European Photon Imaging Camera (EPIC), respectively – have a spectral resolution of $R \approx 50$ at 6 keV, which is typical for Si-based CCDs at this energy.

Higher spectral resolutions for soft X-rays can be reached using microbolometers (called *microcalorimeters* in the context of X-rays) and superconducting tunnel junctions (STJs). The principles and properties of these two detector types were described in Section 5.3. As pointed out in Section 5.3, these detectors require very low operating temperatures. Achieving and maintaining these very low temperatures in space observatories is an even greater technical challenge than the use of such detectors on the ground. A first attempt to use a microcalorimeter array for the 0.5–12 keV X-ray range in space was carried out on the Japanese X-ray satellite Suzaku (Astro-E2). Its TES microcalorimeter array, which had been produced by NASA-GSFC in the United States, had a spectral resolution of $R \approx 920$ at 6 keV. Its operating temperature of 0.06 K was maintained with a four-stage cooling system that included slowly melting solid neon (at 17 K), boiling helium (at 1.3 K), and an adiabatic demagnetization refrigerator. After an earlier version of the experiment had been lost because of a malfunctioning spacecraft, Suzaku was launched successfully in 2005. However, a thermal short between the helium tank and the neon tank resulted in a complete loss of the helium coolant after only a few weeks of operation, which shut down the calorimeter array. Nevertheless, Suzaku has shown that very-low-temperature detectors can be deployed in space; plans are under way to use the Suzaku detector again in the future Astro-H mission (see Chapter 10).

STJs have been used successfully in laboratory X-ray spectroscopy, but not yet in astronomy. However, they are considered for future missions, which will be discussed in Chapter 10.

Si-based CCDs have been used for photon energies up to about 15 keV. At higher energies, semiconductor materials with a higher atomic weight are required. Moreover, because of a smaller effective absorption cross section, larger detectors are needed. The NuStar mission (see Section 7.1.3), designed for the 5–80 keV range, uses CdZnTe detector arrays. Their predicted spectral resolution is about $R = 10$ at 5 keV and about $R = 60$ at 60 keV.

Standard detectors for hard X-rays and gamma rays are germanium diodes. Depending on the atomic weight of the material, at about 20 keV Compton scattering starts to become the dominant interaction of energetic photons with matter. As in the case of the photoeffect, at high energies the Compton scattering results in energetic electrons, which in semiconductors lose their energy by producing conduction electrons. Thus, if fully absorbed, current pulses proportional to the photon energy are produced. Germanium diodes can be used up to

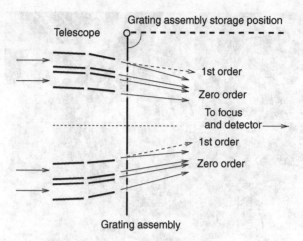

Figure 7.7. Optical layout of the slitless transmission-grating spectrometer of the Chandra X-ray observatory.

the nominal upper limit of the X-ray range at 500 keV. The spectral resolutions in the X-ray range reaches about $R \approx 200$.

7.4 X-Ray Grating Spectrometers

As noted earlier, the most important X-ray observatories that are in operation at the time of this writing are the NASA facility Chandra and the ESA satellite XMM-Newton. Both these facilities are equipped with gratings for medium resolution spectroscopy, and in both cases slitless spectrometers are used. However, the detailed design of the two instruments is quite different.

Figure 7.7 shows the layout of the grating spectrometer of the Chandra observatory. The imaging optics of Chandra consists of four nested type I Wolter telescopes with a common focus. Into the convergent beams behind each of the four telescopes transmission gratings can be placed by means of a hinge mechanism (as indicated in Figure 7.7). Because of the annular geometry of the four output beams of the telescopes, for each telescope the grating surface consists of a mosaic of small aligned gratings, which are arranged in four rings, as shown in Figure 7.8. Chandra is equipped with two such grating assemblies, which can be used alternatively. One assembly (denoted High-Energy Transmission Grating Spectrometer, HETGS) is optimized for the 0.15 nm–2.4 nm (0.5 keV– 8 keV) range, whereas the second grating assembly (Low-Energy Transmission Grating Spectrometer, LETGS) is used for wavelengths 0.6 nm–18 nm. The

Figure 7.8. The two transmission grating assemblies of the Chandra X-ray observatory in their storage position. Image courtesy NASA/CXS/SAO.

HETGS mosaic actually contains two different types of gratings with different line densities and slightly different groove orientations. Thus, the HETGS simultaneously produces two spectra. To minimize the optical aberrations, the individual gratings and the detectors are located on curved surfaces. The detectors are CCDs and microchannel arrays. Because the gratings are placed close to the telescope optics, the optical arrangement corresponds essentially to an objective-grating design (see Section 4.3). However, because of the small field of view of the Wolter telescopes, normally only single objects are observed.

The Chandra transmission gratings consist of fine gold bars or wires on a thin (for X-rays, transparent) plastic substrate. Because the detectors cover more than a decade of wavelengths, several orders are imaged on the detector arrays simultaneously. The orders are separated using the photon-energy resolution of the CCD detectors. Thus, in a sense, the Chandra spectrometers are "cross-dispersed." The spectral resolutions of the Chandra grating spectra are in the range $125 < R < 3{,}600$, and depend on the wavelength. Some examples of Chandra spectra are reproduced in Figure 7.9. For additional examples and more information on the Chandra spectroscopic mode, see Paerels and Kahn (2003).

Compared with Chandra, the ESA facility XMM-Newton has a larger effective aperture area but less angular resolution. The high throughput of XMM-Newton is achieved by combining three co-aligned telescope systems, which in

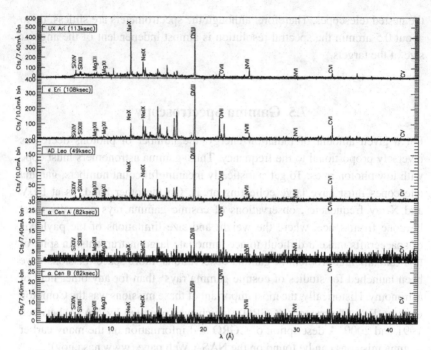

Figure 7.9. Example of X-ray spectra obtained with the low-energy transmission grating spectrometer of the Chandra X-ray observatory. All objects are galactic stars with cool atmospheres and hot circumstellar matter, giving rise to the conspicuous X-ray emission-line spectra. From Ness et al. (2002).

each case consists of fifty-eight nested Wolter type I telescopes. Although one of the three telescope assemblies is used for imaging only, two telescopes send their light into grating spectrographs. Each of these spectrographs contains 182 grazing-incidence reflection gratings, which intercept about 40 percent of the radiation (the rest being sent to an imaging detector). The reflection gratings and the spectral CCD detectors are arranged on Rowland circles. Because of the grazing incidence design, the planes of the individual reflection gratings are almost perpendicular to the circle's circumferences. A schematic optical layout of the XMM-Newton spectrometers, as well as examples of spectra obtained with these instruments, can again be found in the review paper by Paerels and Kahn (2003). (This article also cites literature with additional information.)

The XMM-Newton spectroscopic mode has been designed for the wavelength range 0.5–3.5 nm (0.35 keV–2.5 keV). The orders (1 and 2) are again separated using the energy resolution of the CCD detector. The maximal spectral resolution is about $R = 580$. The optical parameters of the XMM-Newton spectrometers were designed to match the relatively poor angular resolution of the nested telescopes. Therefore, although the spectrometers are slitless, up to about 0.5 arcmin the spectral resolution is almost independent of the angular size of the targets.

7.5 Gamma Spectroscopy

For a given amount of radiation energy, the number of photons decreases inversely proportional to the frequency. Thus, gamma astronomers must cope with low photon rates. To get statistically meaningful count numbers, gamma telescopes must have large collecting areas. On the other hand, as at EUV and X-ray frequencies, observations of cosmic gamma rays normally must be done from space, where the weight and size limitations of the payloads of spacecrafts make it difficult to accommodate large instruments. In spite of this fundamental problem of gamma observations, more space missions have been launched for studies of cosmic gamma rays[1] than for any other field of astronomy. Historically, the most important of these missions was the Compton Gamma Ray Observatory (CGRO), which was operated by NASA between 1991 and 2000. A description of CGRO and information on the many earlier gamma missions can be found on the NASA Web page (www.nasa.gov).

Because optical methods are not feasible at gamma-ray photon energies, the spectroscopy in this range depends fully on photon-energy-sensitive detection techniques. The corresponding detectors are based either on the Compton effect in solids or on the production of electron-positron pairs.

7.5.1 Diodes and Scintillation Detectors

As noted in Section 7.3, for the spectroscopy of hard X-rays germanium diode detectors are often employed. Such devices also provide a good spectral resolution for gamma rays up to about 10 MeV. Among the presently active X-ray observatories, the INTEGRAL satellite of ESA (www.isdc.unige.ch/integral) makes use of this technique. INTEGRAL includes an array of cooled germanium detectors for spectroscopy in the photon-energy range 18 keV–8 MeV. The energy resolution element is about 2.2 keV, corresponding to a spectral resolution $R \approx 600$ at 1.3 MeV.

At higher energies (up to about 30 MeV), the scintillation effect caused by Compton electrons in crystalline solids is used to detect and measure gamma photons. The scintillation light flashes are recorded with optical detectors. The

---

[1] A NASA statistic lists more than eighty gamma missions worldwide between 1961 and 2010.

energies of the individual gamma photons are derived from the intensity of the corresponding flashes. The most commonly used scintillators are NaI or CsI crystals, doped with Tl or Na impurities. Scintillation detectors are sensitive over a large energy range, but their spectral resolution (typically of the order $R \approx 20$) is modest.

Like semiconductors, scintillation detectors are also sensitive to energetic charged particles. To distinguish between scintillation flashes caused by gamma photons and those originating from charged particles, scintillation detectors are surrounded by sheets of specific particle detectors (such as organic plastic scintillators) that record the charged particles, but that interact only weakly with high-energy photons. Using these "active shielding" devices and anti-coincidence techniques, particle events can be eliminated efficiently from gamma observations with scintillation detectors.

7.5.2 Pair-Production Detectors

At energies >30 MeV, electron–positron pair production becomes the dominant interaction of gamma rays with matter. In principle, any photon can be converted into an electron–positron pair if its energy is larger than the combined rest-mass energy of the two new particles (which is 1.022 MeV). However, because photons always carry momentum, pair production can occur only if, in addition to the energy, the momentum is conserved as well. This requires the presence of an additional particle that can pick up momentum. In principle, this can be another photon. However, a heavy atomic nucleus is much more efficient. Therefore, pair-production detectors typically contain sheets of high-atomic-weight solids. Once electron–positron pairs have been generated, they are tracked and detected using the well-developed standard methods of particle physics.

In the past, spark chambers were used to record cosmic gamma rays by means of their pair-production capability. Present-day high-energy gamma detectors, such as the Large Area Telescope (LAT) on NASA's Fermi gamma-ray observatory (http://fermi.gsfc.nasa.gov), use alternating layers of metal sheets (where electron–positron pairs are produced) and semiconductors (where the charged particles are tracked). Fermi uses fifteen planes of tungsten sheets, which are interleaved with silicon-strip detectors. This assembly is followed by a calorimeter array, which consists of crystal scintillators. The calorimeters measure the total energy of the electron–positron pairs (and thus, indirectly, the energy of the incident gamma photons). The scintillators are read out using light-sensitive PIN diodes. Silicon-strip detectors (today's standard devices for

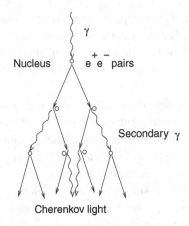

Figure 7.10. The production of air showers and Cherenkov radiation in the Earth's atmosphere by energetic gamma photons.

precisely measuring positions of particle beams) are used to track the particles and to derive the position on the sky from where the gamma photons originated.

The spectral resolution of LAT is given by the energy resolution of the calorimeter, and amounts to $R \approx 25$ at 5 GeV and $R \approx 50$ at 200 GeV. A detailed description of the LAT detector system of the Fermi observatory has been given by Atwood et al. (2009).

At very high gamma energies, the attainable spectral resolution often becomes limited by the statistical error of the photon counts. A simple estimate shows that at a photon energy of 10^{14} eV, even for a source that (expressed in W m^{-2}) is as bright as the brightest stars on the sky, we would receive fewer than ten photons per hour, per square meter and per frequency decade. In practice, cosmic gamma sources are much fainter. Therefore, at TeV energies the detector surfaces of space-based gamma observatories (about 2.5 m^2 in the case of LAT, much less in other gamma observatories) are simply too small to collect enough photons for spectroscopic applications. For this reason, at the highest energies (10^{11} to 10^{14} eV) gamma astronomy makes use of the natural electron–positron pair production by gamma photons in the Earth's atmosphere. As indicated schematically in Figure 7.10, in the Coulomb field of the nucleus of an air atom a TeV gamma photon from space can generate an ultra-relativistic electron–positron pair. Interacting with other atomic nuclei, these particles lose energy by emitting bremsstrahlung. Because of the high particle energies, the bremsstrahlung photons are energetic enough to generate secondary relativistic electron–positron pairs. This process repeats itself until the energies of the secondary photons become too low for efficient further pair production. As a result of the pair-production cascade, each TeV gamma photon results in an air shower of charged particles in the upper atmosphere, typically about 10 km above sea level.

If the initial photon energy is larger than about 10^{13} eV, the shower of charged particles can extend to the Earth's surface, where it can be measured using particle detectors. However, gamma photons of such high energies are rare, as (because of photon–photon reactions with the intergalactic sky background) the intergalactic space is essentially opaque for photons with energies above about 20 TeV, and because few galactic sources reach such high photon energies. On the other hand, the particle showers in the high atmosphere can be observed on the ground indirectly by the visual Cherenkov radiation that is emitted by the ultra-relativistic charged particles. The highly energetic electrons and positrons propagate with velocities that are just below the vacuum velocity of light, but that are (slightly) higher than the phase velocity of light in the atmospheric air. As a result, these particles produce Cherenkov radiation. The highly relativistic velocities of the particles cause this Cherenkov radiation to be emitted essentially into the forward direction of the particles (and of the original gamma photon). A calculation shows that the Cherenkov photons are emitted into a narrow cone (with an opening angle of about 1 degree) around the propagation direction of the original gamma photon. This cone forms a light spot on the ground (called the *Cherenkov light pool*) with a diameter of about 250 m. The Cherenkov radiation can be observed by pointing optical telescopes toward the gamma ray source. As a result of the cascade process described earlier, the total number of Cherenkov photons is about proportional to the initial gamma photon energy. This total number of photons (and thus the gamma photon energy) can be measured if the area of the light pool is fully covered by optical telescopes. However, if only part of the pool is covered, the total number can still be inferred if the size of the light pool and the covering factor are known.

Although the Cherenkov photons are emitted all along the air shower at different heights and times, they arrive almost simultaneously (within a few nanoseconds) because the particles and the light propagate with almost the same velocity. Thus, for each gamma photon the optical telescopes see a very short visual light flash, which can be easily separated from other radiation sources on the sky.

Air showers and Cherenkov light are also produced by relativistic cosmic ray particles (such as protons). However, in this case additional hadronic reactions take place in the atmosphere. Therefore, cosmic ray air showers are broader and less symmetric than the showers produced by gamma photons. By means of stereoscopic observations with several telescopes, the geometric structure of the

Figure 7.11. One of the (at the time of writing) four optical telescopes of the High-Energy Stereoscopic System (H.E.S.S.) for the detection and spectroscopy of TeV gamma rays. The mirror surface consists of 382 adjustable segments. Each segment has a diameter of 60 cm. The prime-focus camera (at the left) has a 5-degree FOV and consists of 960 photomultipliers. Image courtesy H.E.S.S collaboration.

showers can be determined and gamma photon air showers can be distinguished reliably from those caused by cosmic rays.

From the principles described previously, it follows that measurements of cosmic gamma rays by means of the air-shower technique require arrays of optical telescopes distributed over the area of the Cherenkov light pool. The individual telescopes need not to be of high optical quality, but their total surface area should cover a significant fraction of the light pool to record a representative part of the Cherenkov photons. Normally, arrays of large segmented mirrors are used for this purpose. The detectors have to be able to image a field of a few degrees with modest angular resolution, but with a high time resolution. An example of an optical telescope used in an air-shower gamma-ray observatory is reproduced in Figure 7.11. The figure shows part of the High Energy Stereoscopic System (H.E.S.S.), a Cherenkov air-shower gamma telescope that is operated by an international consortium at a site in western Namibia. H.E.S.S. is designed for the detection of gamma photons in the range 0.1–100 TeV. It has an energy (and spectral) resolution of $R = E/\Delta E \leq 10$.

This value is less impressive than the resolutions that can be achieved at lower gamma energies. However, because H.E.S.S. can record spectra over photon energy range of three decades, its spectra can provide unique information on the continua of gamma sources at very high energies. A comprehensive description of the H.E.S.S. system, and of the Cherenkov air-shower technique in general, has been given by Völk and Bernlöhr (2009). A more general description of the field of ground-based TeV astronomy can be found in a recent review by Hinton and Hofmann (2009).

8

Spectroscopy at Radio Wavelengths

Radio astronomers typically use coherent receivers, which record and analyze the electromagnetic radiation directly. As was pointed out in Section 1.4, this facilitates spectroscopy at radio wavelengths. Making use of electronic frequency filters or of natural resonance effects, receivers can be built that are sensitive to a narrow frequency range only. If such receivers contain electronic components for which the parameters can be varied, the receivers can be "tuned" to specific frequencies. In this case, spectra can be be obtained by tuning narrowband receivers within a certain wavelength range, or by combining receivers that are tuned to different frequencies. As described in Section 1.4, the first radio spectra were obtained in this way, and commercially available radio spectrometers still use this technique. However, if spectra are obtained by tuning a receiver, the different frequencies are measured sequentially and the duty cycle for a given frequency is inversely proportional to the spectral resolution. Therefore, such spectrometers are inefficient and not well suited for the low radiation levels of faint astronomical radio sources. To study the very low signal levels from cosmic sources, more efficient methods must be used, which allow us to record many frequencies simultaneously. Some of these methods are described in the following sections.

The aim of this chapter is to provide an introduction to the technical background of the different types of radio spectrometers at a level that an observer needs to select the optimal instrument for a given task and to assess the potentials and the limitations of the different methods. The scope of this book and the available space do not permit discussion of the complex physical and engineering details of radio-wave spectrometers. Moreover, because of the fast progress of microelectronics and digital methods, radio spectroscopy is developing very rapidly and new techniques are being introduced at a fast pace. Readers interested in technical details and in a timely and comprehensive general introduction to the methods of radio astronomy will find an excellent

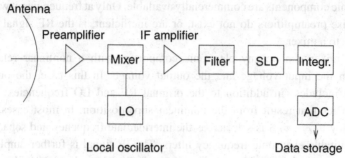

Figure 8.1. Schematic layout of a simple receiver for measuring the radio flux of astronomical sources. The individual components are explained in the text.

treatise of this field in the book *Tools of Radio Astronomy* by Wilson et al. (2009). Valuable up-to-date information on the currently available radio spectrometers and on new developments can also be found on the Web pages of all major radio observatories.

8.1 Detection of Radio Waves from Space

Radio waves from cosmic objects are received with large reflecting telescopes, telescope arrays, or (at low frequencies) arrays of dipole antennae. In radio astronomy, all these devices are called *antennae*. To measure the intensity of a radio source, it is sufficient in principle to amplify the electric high-frequency signal at the output of an antenna, to rectify the resulting high-frequency AC voltage, and to record the resulting DC signal.

For various reasons, such simple linear receivers are not practicable for radio astronomy. One reason is the high amplification factor (typically $> 10^8$) that is required for the low flux levels of astronomical sources. Such high amplification factors are difficult to achieve in one stage without feedback and interference effects. Moreover, signal processing at the frequencies emitted by astronomical objects is often inconvenient. Therefore, radio astronomers normally use (super)heterodyne receivers, in which the observed signal is mixed with that of a local oscillator (LO) and converted to an intermediate frequency (IF) signal. The principal layout of a simple radioastronomical heterodyne receiver is outlined in Figure 8.1. Normally, the signal from the antenna is sent through wave guides to a low-noise preamplifier, which brings the electric signal to a level at which it can be processed without picking up significant additional noise. The amplified radiofrequency (RF) signal is then fed into a mixer, which combines the RF signal with the signal of the local oscillator. By choosing a suitable LO frequency, an intermediate frequency can be achieved for which electronic components are commercially available. Only at frequencies at which low-noise preamplifiers do not exist, or are inefficient, is the RF signal sent directly to a mixer.

The mixer can be any electronic component with a nonlinear relation between the input voltage and the output voltage. In this case, the output voltage includes (in addition to the original RF and LO frequencies) new frequencies that result from the nonlinear superposition. In most cases, the frequency $\nu_{RF} - \nu_{LO}$ is selected as the intermediate frequency and separated using suitable electronic frequency filters. The IF signal is further amplified in one or several steps. If needed, or desired, the IF band can be tailored to an arbitrary specific form by additional mixing with different local oscillators. By filtering out all frequencies above a certain limiting value and mixing with an LO frequency corresponding to this limit, it is possible, for example, to generate from any original RF frequency interval an IF-frequency band that extends from $\nu = 0$ to a value ν_{max}.

The RF frequencies $\nu_{LO} + \nu_{IF}$ and $\nu_{LO} - \nu_{IF}$ differ from the LO frequency by the same absolute value. Hence they result in the same intermediate frequency. Thus, a priori, each frequency ν_{IF} corresponds to two different values of ν_{RF}. This is no problem for continuum measurements, as the two frequencies normally are of the same order. However, this ambiguity can lead to problems in the identification of spectral lines. To avoid this, one of the two frequencies can be eliminated by filtering the RF signal. This is done in single-sideband receivers. However, this filtering tends to increase the noise. Therefore, even for line spectroscopy, double-sideband receivers are often used, which record both frequencies. In this case, the frequency ambiguity must be resolved using physical information, such as the expected frequencies of the spectral lines.

After the IF (and the corresponding RF frequency to which the receiver is sensitive) has been selected using a final IF band-pass filter, the IF voltage is converted into a DC voltage, integrated over a suitable time interval, and converted to a digital signal by means of an analog-to-digital converter. For the rectification of the high-frequency signal, normally a square-law detector (SLD) is used. An SLD has a DC output voltage that is proportional to the square of the AC input voltage. Because the power of an electromagnetic wave is proportional to the square of the amplitude of the electric field, the output of the SLD provides a direct measure of the radiation power of the observed source in the selected wavelength band.

If all gain and attenuation factors of the electronic components (including the antenna) are known, the flux of the observed source can be calculated from the squared output voltage. However, these parameters may change with time. Therefore, as at other wavelengths, during most observations radio receivers are calibrated using standard sources on the sky. The flux values of these standard sources are calibrated by a comparison with the radiation of a low-temperature black body (which can be calculated precisely from Planck's law) or by comparison with the AC noise power P of an Ohmic resistor, which is given by the Johnson-Nyquist relation $P = kTb$, where b is the frequency band width of the instrument and T is the temperature of the resistor.

The first part of the typical receiver of Figure 8.1, which includes the preamplifier and the mixer, is called the *front end*. The most common preamplifiers at present-day radio telescopes are cooled low-noise transistors, such as high-electron-mobility transistors or HEMTs (see Section 5.2 of Wilson et al., 2009). Modern mixers often make use of the highly nonlinear voltage response of superconducting tunnel junctions (see Section 5.3 of this book). At very high frequencies, SIS tunnel junctions or hot-electron bolometers (see Chapter 9) are used as mixers without preamplifiers.

The receiver sections following the (first) mixer are called the *back end*. Radio spectrometers are always part of this back end. Therefore, in the following the front ends will not be discussed further. Detailed information on the front ends, which are used currently at large radio telescopes, can again be found in the textbooks of radio astronomy.

8.2 Filter Banks

The basic receiver of Figure 8.1 permits measuring the flux in the wavelength band defined by the frequency filter of the IF circuit. A simple extension for spectroscopic applications is shown in Figure 8.2. In this layout, the single band-pass filter is replaced by a series of n frequency filters, which are arranged in parallel. The input for all filters is the amplified IF signal. Ideally, the frequency filters have boxlike transmission functions, which cover the frequency range of the input signal without gaps. Each filter is followed by its own square-law detector, integrator, and analog-to-digital converter. The n channels are read out by a computer, which (using the known LO frequencies) converts the IF frequencies into the original frequency scale of the source.

If the n individual channels are identical (apart from their central frequency), their output gives directly the power per filter band as a function of the frequency. Thus, we get a spectrum with n spectral elements and a spectral resolution that is determined by the bandwidth of the individual filters. Because at radio wavelengths photon noise is normally negligible (as $h\nu$ is small), and because the incident waves can be amplified before being rectified, the S/N of the individual spectral elements does not depend on n or the spectral resolution. This situation is quite different from the visual and high-energy spectral ranges,

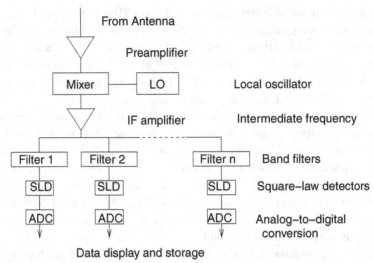

Figure 8.2. Schematic layout of a filter-bank spectrometer.

where for a given object flux and integration time the S/N decreases with the resolution.

Spectrometers based on filter banks have been standard instruments for many years, and they are still used for certain applications. Although such spectrometers are relatively simple, they have operational drawbacks. One problem is that differential variations or drifts between the individual channels tend to cause spurious spectral effects. Differential drifts are much more likely to occur in discrete electronic components of filter banks than in the monolithic array detectors that are used in optical astronomy. Another drawback is the fixed spectral resolution defined by the electronic hardware. Finally, filter banks with many channels require large numbers of well-tuned electronic components, which results in a major technical and financial effort.

8.3 Fast Fourier-Transform Spectrometers

The filter-bank spectrometer, which was described in the preceeding section, determines the spectral energy distribution $S(\nu)$ of an electric input signal $v(t)$ by measuring the power in adjoining discrete frequency bands. From the theory of signal processing, it is clear that (at least in principle) $S(\nu)$ can be derived more directly by making use of the fact that $v(t)$ and $S(\nu)$ are (apart from known factors) Fourier pairs. In detail, we have

$$P(\nu) = \left| \int_{-\infty}^{\infty} v(t) e^{-2i\pi \nu t} dt \right|^2 , \tag{8.1}$$

From mixer

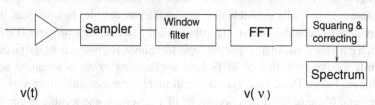

Figure 8.3. The principle of an FFT spectrometer.

where the function $P(v)$ is the observed energy as a function of frequency. Although its dimension is that of an energy, $P(v)$ generally is called the *power spectrum* of $v(t)$. The spectral power $S(v)$ obviously can be derived by dividing $P(v)$ by the effective integration time of a measurement.

Following the definition of a Fourier transform, Equation 8.1 requires a time integration between $-\infty$ and $+\infty$. Only an infinite integration interval can ensure that all possible frequencies are included. In practice, only finite integration times (and measurements of finite frequency ranges) are feasible. The theory shows that the function $P(v)$ becomes distorted if derived from a finite approximation of the Fourier integral. Certain frequencies are suppressed or represented incorrectly. The distortion can be reduced by convolving the signal with a *window function* instead of simply cutting off the time interval. The theory of Fourier transforms shows that, in this case, we get as a result the product of the correct Fourier transform of $v(t)$ and the Fourier transform of the window. Therefore, if a suitable window function is applied, its effect can be corrected numerically.

The technical realization of a spectrometer that is based on a Fourier transform of the input signal is outlined in Figure 8.3. After the RF signal has been transformed into an IF signal of a suitable bandwidth and shape, it is converted to a digital signal by sampling the IF voltage at constant time intervals. After applying a window function, a Fourier transform converts the digital signal $v(t)$ into a spectrum $v(v)$. This spectrum is numerically corrected and squared to get the flux spectrum $S(v)$.

Calculating the Fourier transform obviously is the most straightforward method to derive a spectrum of radio radiation. However, because the radio waves must be sampled in real time, this type of spectrometer requires fast electronic components and fast processors to calculate the Fourier transforms. According to Nyquist's sampling theorem, a wave must be sampled at least twice during one oscillation period. Thus, for a typical IF band extending up to 1 GHz, the sampling frequency must be at least 2 GHz, and the Fourier transform calculation must be carried out with the same high data rate. Analog-to-digital converters and digital components that can manage these high speeds became commercially available only during the past decade. To save time, the numerical calculations use an algorithm called *fast Fourier transform* (FFT), which gave the name to this spectroscopic technique. High-speed digital circuit boards for the execution of FFTs have applications in many scientific and technical fields. Thus, such boards are available commercially from various suppliers. More detailed descriptions of FFT spectroscopic technique can be found in papers by Klein et al. (2006a and 2006b). Examples of high-resolution line-profile observations obtained with an FFT spectrometer are reproduced in Figure 8.4.

Operationally, FFT spectrometers have important advantages. Because they use commercially produced standard digital components, they can be easily adapted to different applications and requirements, and complete systems are available from several manufacturers of high-frequency scientific equipment. With the present-day digital components, spectrometers with several 10^4 frequency channels and with a high dynamical range can be realized. Because most of the signal processing occurs in digital form, FFT spectrometers are efficient and stable. In contrast to the types of spectrometers discussed in the next sections, the absolute power of a source is measured directly without the need of a calibration with the total flux. A drawback of the presently existing FFT spectrometers are their (so far) limited spectral bandwidths. With the presently available FFT electronic boards, the spectral resolution is limited to about $R \leq 3 \times 10^4$.

8.4 Autocorrelation Techniques

Also based on Fourier transforms are the *autocorrelation spectrometers*. They make use of the fact that according to the autocorrelation (or Wiener-Khinchine) theorem, the spectrum $S(\nu)$ of a wave is equal to the Fourier transform of its autocorrelation function R. Thus, if $R(\tau)$ is the autocorrelation function, we can write

$$S(\nu) = \int_{-\infty}^{\infty} R(\tau)e^{-2i\pi\nu\tau}d\tau, \qquad (8.2)$$

where $R(\tau)$ is given by

$$R(\tau) = \int_{-\infty}^{\infty} v(t+\tau)v(t)dt. \qquad (8.3)$$

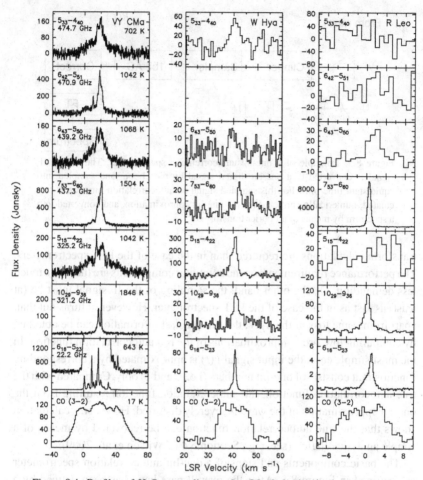

Figure 8.4. Profiles of H_2O maser lines and a CO (3–2) line in the spectra of two red giant stars and a red supergiant star. The profiles of the 321.2–474.7 GHz lines were observed with an FFT spectrometer at the 12-m APEX telescope at Chajnantor, Chile. For the 22.2 GHz line, the Effelsberg 100-m telescope in Germany with an autocorrelation spectrometer was used. The profiles of the mm-range lines are relatively noisy, as the atmospheric H_2O lines make observations at this wavelength very difficult. These observations could be successful only due to the dry high-altitude (5100 m) site of the APEX telescope. From Menten et al. (2008).

Compared with the direct Fourier transform of the FFT spectrometer, the autocorrelation spectrometers require two integrations. On first glance, this seems to be a unnecessary complication. However, because the autocorrelation function can be integrated and averaged in time, fewer and less fast Fourier

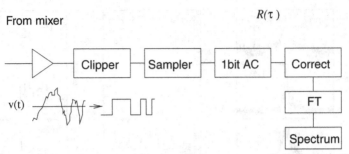

Figure 8.5. Principle of a one-bit autocorrelation spectrometer. The analog IF signal is converted to a one-bit digital signal by the clipper and sampled into equidistant time bins. For this sampled signal an autocorrelation function is calculated, numerically corrected for the one-bit approximation, and converted into a spectrum by means of a Fourier transform.

transform calculations are required than in the case of the FFT spectrometers. The performance requirements on the computer components are therefore much less demanding. On the other hand, the sampling of the IF signal must be (at least) as fast as in the case of the FFT spectrometer. However, it turns out that, apart from some loss of the S/N, the full spectral information can be obtained by computing low-order approximations of the full autocorrelation function. In the most simple case, the input signal $v(t)$ is approximated by a corresponding function that consists of one-bit numbers (i.e., 0 and 1) only. Compared with the true autocorrelation function, the autocorrelation function resulting from the one-bit approximation of the wave is severely distorted. However, a calculation shows that the true autocorrelation function can be recovered by means of a simple analytic relation (see, e.g., Section 5.4 of Wilson et al., 2009).

The basic components and steps of a one-bit autocorrelation spectrometer are outlined in Figure 8.5. After the overall pass band has been defined by a suitable filter, the analog IF signal is converted into a digital signal by means of a *clipper*. In the one-bit case, the clipper converts the voltage signal into a time series of the numbers 0 or 1. This can be done by assigning the number 1 to all voltage values > 0 and the number 0 to all values < 0 (as indicated in Figure 8.5). The resulting digital signal is sampled at equidistant time intervals. The minimum sampling frequency again is given by Nyquist's sampling theorem. However, a better S/N can be achieved with a higher sampling rate. In the next step, the autocorrelation function is calculated by multiplying the one-bit $v(t)$ function with values $v(t + \tau)$. In practice, this is achieved by a multiplication with values $v(t)$ that are shifted in time by multiples of the sampling interval $\Delta \tau$. All resulting products are added up to approximate the integration of Equation 8.3. Because of the one-bit numbers, these multiplications are particularly simple and can be done fast with primitive digital electronic components. After integrating and averaging the autocorrelation function over an appropriate time period, the function is corrected for the clipping and converted to the spectrum $S(\nu)$ by a final Fourier transform.

If properly corrected, the clipping does not introduce systematic errors into the final spectrum. However, it reduces the S/N and it erases all information on the original absolute signal strength. The absolute power can be recovered by integrating over the spectrum and by comparing the result with the total power in the corresponding frequency band. To derive the total power, an additional independent flux measurement must be carried out.

The loss of S/N resulting from the clipping procedure depends on the rate at which the clipped signal is sampled. If the sampling rate corresponds just to the Nyquist condition, the S/N is reduced by a factor of 0.64. With a sampling rate twice as high, a factor of 0.74 is reached. Higher values can be obtained by converting the analog input signal into a digital signal with more than one bit. A three-bit signal (which can still be generated and processed with relatively simple electronic components) and a sampling rate twice the Nyquist value result in an S/N that is 0.94 times the ideal value. Therefore, many modern AC spectrometers use two-bit or three-bit electronics.

Like the FFT spectrometers, the autocorrelation instruments are stable devices containing simple and reliable electronic components. At present, they are among the most frequently used types of radio spectrometers. An example of a radio spectrum obtained with an autocorrelation spectrometer is the profile of the 22.2 GHz H_2O line in Figure 8.4.

8.5 Cross-Correlation Spectroscopy with Arrays

As pointed out in Chapter 4, the angular resolution of a telescope depends on the ratio D_T/λ, where D_T is the aperture of the telescope and λ is the wavelength. Compared with the visual light, the wavelengths used in radio astronomy are typically larger by factors between 10^3 and 10^7. Thus, to reach the same angular resolution, radio instruments must have sizes of the order of kilometers or more. Because such sizes cannot be achieved with single telescopes, radio astronomers often use antenna arrays. These arrays make use of the fact that (apart from known factors) the brightness distribution of an object on the sky and the combined signal of two antennae as a function of the projected distance of the antennae are Fourier pairs. Thus, the brightness distribution can be calculated if the combined signal is measured as a function of the antenna distance. The technical details of this interferometric imaging technique can again be found in the textbooks of radio astronomy.

Although the Fourier transform of a brightness distribution can be obtained directly by adding the signals of two antennae, a superior S/N can be achieved by calculating the Fourier transform of the cross correlation between the individual signals. For this purpose, one determines the cross-correlation function, which (analogous to Equation 8.3) is defined as

$$C(\tau) = \int_{-\infty}^{\infty} v_i(t + \tau)v_j(t)\mathrm{d}t, \qquad (8.4)$$

where v_i and v_j are the output voltages from the antennae i and j. If the two antennae receive radiation from the same source, the cross correlation obviously contains the autocorrelation function of the observed light, which, as explained in Section 8.4, contains information on the spectrum. Therefore, in addition to providing information on the spatial structure of a source, the cross correlation can be used to obtain spectra.

Today, virtually all radio interferometers use the cross-correlation technique. With flexible digital correlators, spectral information often is automatically provided in the standard output products of these instruments. However, depending on the observer's interest, the reduction process normally can be optimized to get a better spectral coverage and resolution at the expense of less spatial information. The corresponding trade-offs are described in the handbooks of the individual arrays. An example of a highly versatile modern antenna array is the international Atacama Large Millimeter/Submillimeter Array (ALMA) in northern Chile. ALMA consists of sixty-six movable antennae with diameters of 12 m or 7 m. A good description (well accessible to the nonspecialist) of the ALMA correlator system and its potential has been given by Baudry (2011).

8.6 Acousto-Optical Instruments

Although FFT and autocorrelation spectrometers are well suited for high resolution spectroscopy, their bandwidths are limited by the speed of their electronic components. An alternative for spectroscopy in broad bands are the *acousto-optical spectrometers* (AOSs), which derive radio spectra by means of acoustic and optic effects. The basic layout of an AOS is outlined in Figure 8.6. In this type of spectrometer, the amplified radio signal is converted into a traveling ultrasonic wave in a solid crystal by means of a piezoelectric transducer (PT). The traveling ultrasonic wave results in density and refractive index variations in the crystal, which follow the wave pattern. If the crystal is transparent and illuminated with a parallel beam of optical light, the crystal acts as a volume phase grating (see Section 3.3.4) for the incident light. Devices that convert

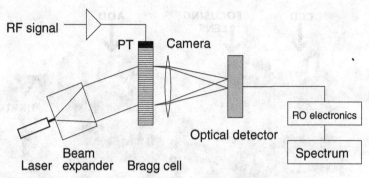

Figure 8.6. Schematic layout of an acousto-optical spectrometer.

an acoustic wave into an optical modulation are called *Bragg cells* or *acousto-optical deflectors* (AODs). In an AOS, the Bragg cell is illuminated with a collimated monochromatic beam from a laser. The optical diffraction pattern produced by the sound wave in the Bragg cell is imaged by an optical camera to a CCD or a similar optical array detector. If the RF signal is a pure sine wave (which corresponds to a radio spectrum consisting of a single spectral line), the refractive index variations in the Bragg cell are strictly periodic, with a period corresponding to the wavelength of the ultrasonic wave. From the discussion of gratings in Chapter 3, it is clear that in this case the diffraction pattern consists (for one order) of a single line. Thus, the pattern on the detector corresponds exactly to the spectrum of the radio radiation. This result also applies to more complex spectra. Thus, the monochromatic optical pattern on the detector gives the spectrum of the observed radio radiation directly. Physically, this can be understood by recalling that the collimated laser beam results in Fraunhofer diffraction at the Bragg cell. Moreover, according to optical theory, the diffraction pattern recorded by a camera is equivalent to a Fourier transform of the camera's entrance pupil (as shown, e.g., in Section 4.2 of Léna et al. (1998)). Thus, if we regard the Bragg cell as the entrance pupil of the camera, we see that the detector observes directly the wavelength spectrum of the wave traveling through the Bragg cell, which (apart from a scale transformation) is that of the radio wave. An exact mathematical derivation of this effect is given, for instance, in Born and Wolf (1987).

Most of the components of AOSs are readily available from suppliers of electronic and optic hardware. However, the performance of an AOS depends critically on the quality and the properties of these components. For the Bragg cells, crystals with a low sound velocity must be selected to obtain an effective grating period that is compatible with the wavelength of the laser light. Moreover, the transducer efficiency must be (as far as possible) independent of the

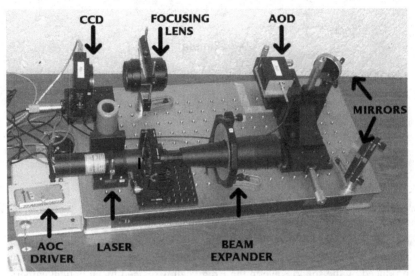

Figure 8.7. Acousto-optical spectrometer built for the Sierra Negra Observatory in Mexico. The functions of the different components were explained in the context of Figure 8.6. A detailed description of the Sierra Negra AOS has been given by Herrera-Martinez et al. (2009). Image courtesy Luis Carrasco, INAOE, Mexico.

frequency, and the response of the Bragg cell must be linear. Finally, because stray light in the system can reduce the S/N significantly, stray light suppression measures must be included. Normally, differences of the polarization properties of the diffracted light and the stray light are used for this purpose.

Descriptions of the technical details of AOSs and their practical performance have been presented, for example, by Schieder et al. (2003) and Herrera-Martinez et al. (2009). An example of a modern AOS is reproduced in Figure 8.7. At present, acousto-optical spectrometers with spectral bandwidths up to 4 GHz are operated successfully at many different telescopes, including the Herschel Space Observatory. For space applications, the AOS has the advantage of a small size and a relatively low power consumption. By choosing suitable Bragg cells and matching optical parameters, a range of different bandwidths and spectral resolutions can be realized. The main application of the AOS technique remains broadband spectroscopy of moderate resolution.

8.7 Chirp-Transform Spectrometers

In the context of waves, the term *chirp* denotes a wave pulse whose frequency varies as a function of time. Chirp technologies were first introduced in high-power radar systems, but soon found many other applications, including astronomical spectroscopy. Chirp-transform spectrometers (CTSs) are based on a special type of Fourier transformation, which is accomplished using fast analog electronic (and acousto-electronic) components. The main application of the CTS technique is spectroscopy at very high frequencies at which (so far) the direct FFT spectrometers cannot be employed.

Mathematically, the chirp-transform follows from the Fourier relation between the input signal $v(t)$ and its frequency spectrum

$$v(\nu) = \int_{-\infty}^{\infty} v(t) e^{-2i\pi\nu t} dt. \tag{8.5}$$

If we have a (chirp) relation between the frequency ν and the time delay τ of the form $\nu = \mu\tau$, we get from Equation 8.5

$$v(\mu\tau) = \int_{-\infty}^{\infty} v(t) e^{-2i\pi\mu\tau t} dt. \tag{8.6}$$

Making use of $2t\tau = t^2 + \tau^2 - (t-\tau)^2$, Equation 8.6 can also be written

$$v(\nu) = \int_{-\infty}^{\infty} v(t) e^{-i\pi\mu[t^2+\tau^2-(t-\tau)^2]} dt \tag{8.7}$$

or

$$v(\nu) = e^{-i\pi\mu\tau^2} \int_{-\infty}^{\infty} [v(t) e^{-i\pi\mu t^2}][e^{i\pi\mu(t-\tau)^2}] dt. \tag{8.8}$$

In a CTS, the two terms in square brackets of Equation 8.8 are generated electronically by analog devices. Both terms require a frequency-dependent time delay. This is accomplished by converting the high-frequency electric wave into an ultrasonic acoustic wave, which is sent through a medium (or over a surface) with a strong frequency dependence of the sound velocity (i.e., a "highly dispersive" medium). The analog multiplications are carried out using mixers. For detailed descriptions of a modern CTS (and references to the earlier literature), see Villanueva and Hartogh (2004).

Chirp-transform spectrometers are used in ground-based and space-based spectroscopy at high radio frequencies and at far-infrared wavelengths. They normally lack the broad frequency range of AOS devices. However, because they contain only electronic components and no optical parts, they are technically simpler and tend to be more stable than AOSs.

9

Special Techniques of the FIR and Submillimeter Range

So far, we have discussed two basic types of spectroscopic techniques. At high photon energies, the observations were based on the detection of individual photons. Their frequency was determined either by measuring their energy or by measuring their wavelength by means of optical effects. At low (radio) frequencies, the electromagnetic waves were directly recorded, and their frequency distribution was derived using electronic methods. As noted in Chapter 3 (Equation 3.42), under thermal equilibrium conditions, the detection of individual photons requires photon energies $h\nu > kT$. Thus, depending on the detector temperature, the transition between photon detection and radio-astronomical methods is expected to take place at FIR or submillimeter wavelengths. In practice, there exists a significant overlap of frequencies at which both types of detection methods can be used, and for which the preferred technique depends on the specific scientific objective. Moreover, in the submillimeter range it is sometimes of advantage to use *bolometers*, which record light indirectly by measuring the heat that is produced when photons are absorbed. Because of the choice of methods, special techniques have been developed for astronomical observations at these wavelengths, and sometimes combinations of radio and optical techniques are employed. A good example of the diversity of methods used in the FIR/submillimeter range are the three spectroscopic instruments of the Herschel Space Observatory (see Figure 9.4), which (as will be described later) use three different techniques.

The purpose of this chapter is to give a brief introduction to the special methods of the FIR/submillimeter range and to discuss their relative advantages and drawbacks for practical observations. Included in this discussion are spectroscopy with bolometers and IR spectroscopy with coherent receivers. IR spectroscopy with infrared-sensitive semiconductors uses the methods described in Chapters 4 through 6. Therefore, these "conventional" techniques will not be discussed again here.

214

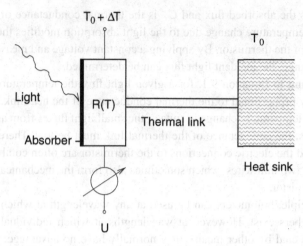

Figure 9.1. Basic components of a single-pixel bolometer.

9.1 Spectroscopy with Bolometers

9.1.1 Basic Physics of Bolometers

As pointed out in Section 5.3, bolometers measure light fluxes by recording the temperature increase resulting from the absorption of light quanta. Section 5.3 described microbolometers, which are used to measure the energies of individual photons at optical and X-ray wavelengths. In the FIR and sub-millimeter ranges, the energies of individual photons cannot be resolved with present-day bolometers. Instead, in that range bolometers measure (similar to array detectors in the visual) the total flux or the spectral flux of the incident light.

The basic principle of a bolometer is shown in Figure 9.1. Light reaching a bolometer is captured by an absorbing layer. The absorber material depends on the wavelength for which the bolometer is designed and on the operating temperature. The active part of the bolometer is a *thermistor*, R(T), which is an electronic component whose electric resistance depends strongly on the temperature. The thermistor is thermally linked to a heat sink of a constant temperature T_0. Normally the heat sink is a boiling liquid, and T_0 corresponds to the boiling point of the liquid. If no light (or other energy) is received by the thermistor, it has the same temperature T_0 as the heat sink. If a constant light flux is received, the thermistor temperature is increased by

$$\Delta T \propto \frac{F}{C_T}, \tag{9.1}$$

where F is the absorbed flux and C_T is the thermal conductance of the heat link. The temperature change due to the light absorption modifies the electric resistance of the thermistor. By applying a constant voltage and measuring the current change, the incident light flux can be determined.

According to Equation 9.1, for a given light flux the temperature change is inversely proportional to the thermal conductance of the heat link. To get a significant temperature change in spite of the small light fluxes from astronomical objects, the conductance of the thermal link must be low. Therefore, the thermal and the electric connections to the thermistor are often combined and realized as two thin wires, which sometimes also form the mechanical support of the thermistor.

In principle, bolometers can be used at any wavelength at which efficient light absorbers exist. However, at wavelengths at which individual photons can be detected by other means, they normally have no advantage. Bolometers always have the disadvantage of being sensitive to thermal backgrounds, though, including the thermal radiation of their own instrumental environment. In astronomy, to avoid such background contributions, bolometers and their immediate environment are cooled to temperatures <1 K. If operated at such low temperatures, bolometers are, at present, the most sensitive detectors for broadband observations in the FIR and submillimeter range. Moreover, using photolithographic or micromechanical techniques, it is possible to produce large-format monolithic arrays of bolometers for these wavelengths.

To reach high S/N values, the thermistor of a bolometer must have a steep dependence of its resistance on the temperature. At present, either semiconductors (such as doped germanium) or superconducting transition edge sensors (TES, described in Section 5.3) are used as thermistors. Examples of current large astronomical bolometer arrays are the LABOCA instrument at the APEX telescope at Chajnantor, Chile, which uses semiconductor thermistors (see Siringo et al., 2009), and SCUBA-2 at the James Clark Maxwell Telescope on Mauna Kea, Hawaii, which uses TES thermistors (see Bintley et al., 2010). Apart from describing the corresponding specific instruments, these two publications also provide more general information on the current astronomical bolometer technologies.

Similar to the optical array detectors, bolometers based on semiconductors and TES are sensitive to natural radioactivity and to cosmic ray particles. The heat input from a charged particle can be large, and discriminating between photons and particles is more difficult in a bolometer than in an optical array detector. Thus, particle absorption is a more serious problem for a bolometer than for a CCD. To avoid a high particle background, the thermistors are made

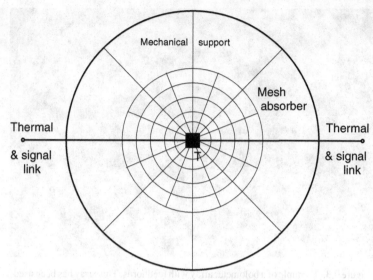

Figure 9.2. Schematic layout of a spiderweb bolometer.

as small as possible, which reduces their cross section to particle absorption. On the other hand, to absorb light, structures of the order of the observed wavelength or larger are required. This condition led to the development of the so-called *spiderweb bolometers*, which have a structure as outlined in Figure 9.2. In the center is the thermistor (T), which is thermally and electrically connected by the two thin wires labeled "thermal and signal link." The thermistor is surrounded and in thermal contact with a mesh of very thin wires. The mesh is sufficiently dense to form an absorbing antenna for the FIR or submillimeter radiation, but its cross section for particle absorption is negligible. Using a properly designed mesh geometry, the refractive index change for the incident waves reaching the mesh can be minimized. In this way, reflection effects at the absorber can be avoided or reduced.

The "spiderwebs" are produced starting from the same type of silicon wafers that are used for integrated circuits. The web wires are converted to silicon nitride (Si_3N_4) and thinly coated with a metal film, while the silicon is removed. Light of very short wavelengths can pass through the mesh, whereas light with wavelengths much larger than the spiderweb structure is not absorbed efficiently. Thus, the design of the mesh also helps to narrow the wavelength range for which the bolometer is sensitive. In this way, the unwanted absorption of thermal backgrounds can be reduced. Instead of the rotationally symmetric

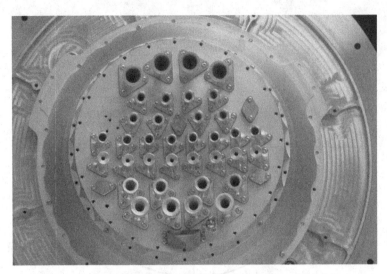

Figure 9.3. Example of a bolometer array with feedhorns. This array has been used on the Planck CMB mapping satellite. Feedhorns of different sizes correspond to different frequency bands. Each feedhorn is followed by a spiderweb bolometer. Image courtesy NASA/ESA.

mesh configuration of Figure 9.2, often rectangular or hexagonal structures are used, which simplifies the construction of large bolometer arrays.

To obtain IR/submillimeter images, either the bolometer arrays are placed directly into the focal plane of a telescope, or each bolometer element is preceded by a *feedhorn* of the type used in the foci of radio telescopes for feeding radio waves into waveguides. An example of an array with feedhorns is reproduced in Figure 9.3. Feedhorns can further reduce reflection effects at bolometers and provide a cleaner coupling to the detector. However, direct imaging on an array of spiderweb bolometers results in a more efficient coverage of the focal plane. Therefore, bolometer arrays with and without feedhorns are used in astronomical spectrometers.

9.1.2 Practical Application to Spectroscopy

As described earlier, bolometers are broadband detectors. To be used for spectroscopy, they must be combined with optical components. To take advantage of the imaging capability of bolometer arrays, normally either Fourier transform optics or Fabry-Perot etalons are used, but there also exist designs for spectrometers combining bolometers with diffraction gratings (see, e.g., Stacey et al., 2004).

FIR/submillimeter Fourier transform and Fabry-Perot spectrometers have the same basic layout as described for shorter wavelengths in Chapter 5. However, because glasses are opaque for FIR radiation, semitransparent mirrors and beam splitters require somewhat different technical solutions. Among the devices used for this purpose at FIR and submillimeter wavelengths are wire grids, thin plastic films, multilayer-coated plastic films, or silicon sheets. A commonly used plastic material is *polyethylene terephthalate*, better known as PET or Mylar. For a discussion of the properties of beam splitters at this wavelength range, see, for example, Homes et al. (2007). Because there exist important applications for far-infrared FT spectroscopy in other sciences, too, efficient beam splitters with a variety of properties are commercially available from several optical suppliers.

Examples for bolometer-FTS combinations are the ground-based SCUBA-2 instrument (which was mentioned in the previous section), and the Spectral and Photometric Imaging Receiver (SPIRE) of the Herschel Space Observatory (Figure 9.4). Both instruments are equipped with spiderweb-type bolometers. SCUBA-2 has two bolometer arrays optimized for atmospheric windows at 450 μm and 850 μm. (The 850 μm window is just short of the wide radio window.) Each of the SCUBA-2 bolometer arrays has 5,120 pixels. The Fourier transform spectrometer of SCUBA-2 was built by a group at the University of Lethbridge in Canada. It provides spectral resolutions in the range $50 < R < 5,000$. A description of this spectrometer was presented by Gom and Naylor (2010).

In the case of SPIRE, two bolometer arrays cover the wavelength range 195 μm to 670 μm. The FTS of SPIRE has a spectral resolution up to $R = 1,000$ at $\lambda = 250$ μm. The details of SPIRE and its FTS have been described in papers by Naylor et al. (2010) and by Griffin et al. (2010).

A spectrometer combining a bolometer and Fabry-Perot etalons has been proposed as a second-generation instrument for the SOFIA stratospheric observatory. This instrument, with the project name SAFIRE, has been designed to work in the wavelength range 145 μm $< \lambda < 650$ μm with a spectral resolution up to $R = 2,000$. (For details see, e.g., Benford et al. (2003).)

9.1.3 Hot Electron Bolometers

A special role in astronomical spectroscopy is played by *hot electron bolometers* (HEBs). HEBs are superconducting devices that make use of the weak interaction of electrons with the phonons in superconducting metals. In

Figure 9.4. The three scientific instruments of the Herschel Space Observatory during the integration at EADS/Astrium. All three instruments include spectroscopic capabilities, but with quite different techniques. PACS contains a grating spectrometers and uses Ge:Ga arrays as detectors. SPIRE employs bolometers and a Fourier transform spectrometer. HIFI is a heterodyne instrument, which uses AOS and autocorrelation spectrometers. Image courtesy Astrium GmbH.

principle, HEBs are small resistors, consisting of extremely thin ($\approx$ nm) metal coatings. Because of the weak electron–phonon coupling in a superconductor, the absorption of photons can result in "hot electrons" with kinetic energies much above the values that would correspond to the temperature of the material. Thus, the absorption of light strongly modifies the electrical resistance of these devices.

HEBs have been used to detect single photons at NIR wavelengths up to 8 µm. Arrays of "nano-HEBs" are being investigated as potential future photon-counting imaging detectors for the mid-infrared, and perhaps even the FIR (see Karasik et al., 2010). The most important present-day astronomical application of HEBs, however, is their use as mixers in high-frequency heterodyne receivers (see, e.g., Tong et al., 2006). HEBs are well suited for this purpose because they are highly nonlinear resistors. Moreover, their small size and the fact that their conductivity depends on a small number of electrons results in a very short response time. Therefore, for frequencies > 1 THz, HEBs are, at present, the most efficient and least noisy mixer elements.

9.2 Heterodyne Spectroscopy at FIR and Submillimeter Wavelengths

The combination of bolometer arrays with FTS or FP optics described previously are efficient tools for FIR/submillimeter imaging spectrometry at low to medium spectral resolution. For high-resolution spectroscopy in the submillimeter range, heterodyne techniques are normally used. Corresponding backends exist at all ground-based large submillimeter telescopes. The most powerful current heterodyne facility for this wavelength range is the cross-correlation spectroscopic system of the ALMA array, which was mentioned in Chapter 8. The 12-m single-dish APEX telescope, which is operated at the same high-altitude site as ALMA, employs a versatile fast Fourier transform spectrometer (denoted XFFTS) as a spectroscopic backend. At the 15-m James Clark Maxwell Telescope on Mauna Kea, Hawaii, an array of autocorrelation spectrometers is used for high-resolution spectroscopy.

High-resolution heterodyne spectroscopic systems are also available at the the currently active space and airborne FIR/submillimeter observatories. In addition to the grating instrument PACS (discussed in Section 4.8) and the SPIRE Fourier transform spectrometer (mentioned in the previous section), the Herschel Space Observatory includes a heterodyne instrument for the far infrared (HIFI), which is designed for the wavelength range 157 μm < λ < 625 μm. HIFI is equipped with two acousto-optical spectrometers and two autocorrelation spectrometers (see, e.g., de Graauw et al., 2010). Spectra illustrating the spectroscopic performance of HIFI are reproduced in Figure 9.5.

On the airplane observatory SOFIA, the echelon spectrometer EXES and the grating spectrometer FIFILS (both discussed in Chapter 4) are supplemented with the heterodyne instrument GREAT. This instrument uses AOS and CTS technologies for very-high-resolution ($10^6 < R < 10^8$) line spectroscopy in selected bands between 60 μm and 240 μm (see Heyminck et al., 2008). A second heterodyne spectrometer for SOFIA is under discussion.

For special applications, heterodyne systems are sometimes used in ground-based high-resolution spectroscopy at mid-IR wavelengths. As pointed out in Chapter 4, at wavelengths longer than 5 μm the crowding of atmospheric spectral lines makes observations from the ground difficult. However, because the telluric lines are narrow ($\Delta\lambda/\lambda \approx 10^{-5}$), with a sufficiently high spectral resolution it is possible to carry out MIR observations in the wavelength gaps between the atmospheric features. The intrinsically high resolution of heterodyne spectrometers obviously is of advantage for this type of work. To achieve convenient IF frequencies, MIR heterodyne systems must use infrared light sources as local oscillators. Normally stabilized lasers are used for this

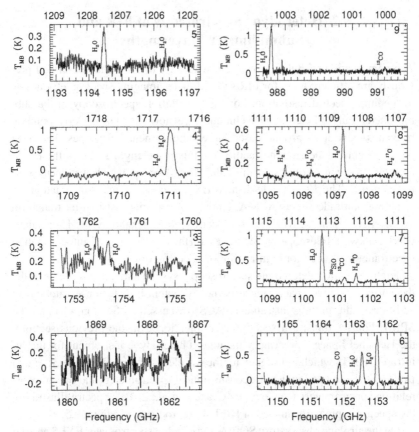

Figure 9.5. FIR/submillimeter emission lines in the spectrum of the oxygen-rich AGB star IK Tauri. These spectra, covering the wavelength range 160 μm to 300 μm, were obtained with the HIFI heterodyne spectrometer of the Herschel Space Observatory. Because the observations were made in double-sideband mode (see Section 8.1), the frequency scale is ambiguous. Therefore, two frequency scales are given for each spectrum. From Justtanont et al. (2012).

purpose. The resulting basic layout of a MIR heterodyne spectrometer is shown in Figure 9.6.

An example of a heterodyne system that has been operated successfully in the MIR is the tunable heterodyne infrared spectrometer (THIS) of the University of Cologne (see Sonnabend et al., 2005). THIS covers the wavelength range 7 μm to 17 μm with a resolution $R = 10^7$. It uses a quantum cascade laser as local oscillator and a mercury-cadmium-telluride photovoltaic detector as a mixer. The spectroscopic backend is an acousto-optical spectrometer. A similar

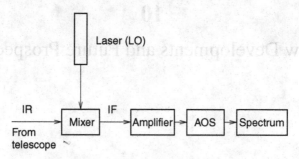

Figure 9.6. Schematic layout of an IR heterodyne spectrometer.

instrument (called HIPWAC) has been developed at the NASA GSFC. The main application of these high-resolution mid-infrared heterodyne systems is in the field of planetary research, but, in principle, such spectrometers can be used for other types of bright astronomical targets, too.

10

New Developments and Future Prospects

Like all fields of science, astronomy is developing rapidly. This also applies to the methods and techniques used in astronomical spectroscopy. It seems safe to predict that many of the instruments, methods, and procedures that have been discussed here will become obsolete during the next decades; and many techniques that are still experimental or unknown at present may become the standard tools of the future. Some new technical opportunities are expected to become available soon. Others will take longer. Some new methods, which are discussed at present, may never become practical. Nevertheless, when planning new scientific programs and future instruments, such new developments and opportunities must be taken into account.

Present-day major astronomical instrumentation projects typically take years or decades to complete. Some large projects are in progress at present, or in advanced planning stages. In these cases, the future scientific opportunities can be assessed with a fair amount of probability. There are other projects and new ideas whose chances of realization are less certain. Moreover, there may be different opinions concerning the potential of some of these ideas. Therefore, the following sections are necessarily more subjective than the earlier chapters of this book.

10.1 Scientific Drivers

Although astronomers do not always fully agree on priorities, a look into the recent relevant publications and reports (such as the 2010 *Decadal Report* of the U.S. National Academy of Sciences, the justification of NASA's Origins program, and the *ASTRONET Science Vision* document produced under the auspices of the European Commission) shows a surprising degree of agreement concerning the most important current research topics in astronomy. All these reports conclude that programs aimed at explaining the phenomena of dark energy and of dark matter (which determine the cosmic dynamics on large scales) should be given high priority. Together with astronomical tests of the theory of the gravitational force (i.e., general relativity), new results on the dark components of the universe are expected to contribute fundamentally to our basic understanding of physics. Other research topics that are generally regarded to be particularly important are the formation and the evolution of the present universe, and the formation of stars, planets, and life. This includes detailed studies of extrasolar planets, of protoplanetary systems, and of the formation of massive black holes in the early universe.

Spectroscopy can contribute significantly to all these topics. However, many of the corresponding observational targets are faint, some require very high S/N data, and part of these studies (e.g., of high-redshift objects and of star-forming regions) depend on sensitive spectroscopy at infrared wavelengths. Therefore, there is rather general agreement that progress in the fields listed earlier needs large space-based and ground-based observatories with sensitive spectrometers covering, in particular, the optical, IR, and radio ranges. Projects for such new facilities have been initiated by different institutions and international cooperations. For some of these projects, much valuable development work has already been invested. Independent of the eventual fate of the corresponding projects, the results of this development work are highly valuable and will sooner or later be applied to the progress in astrophysics.

10.2 New Facilities

10.2.1 TMT

During the last decade of the twentieth century, a new generation of ground-based optical telescopes with apertures between 8 m and 10 m became operational. Characteristic for these highly successful new telescopes was the use of segmented or actively controlled primary mirrors. From the experience with these new instruments, many astronomers concluded that with these technologies significantly larger telescopes could be constructed at acceptable costs. It was also pointed out that such "extremely large telescopes" would be well suited to tackle the unsolved scientific questions listed previously. Therefore, in various places plans for such new large ground-based telescopes have been initiated.

At the time of this writing, the most advanced among these future ground-based projects appears to be the Thirty-Meter Telescope (TMT) that is planned by Californian and Canadian institutions in cooperation with international partners. (For detailed information, see www.tmt.org.) Under optimal conditions, the TMT may achieve first light before the end of the 2010s. The project has a detailed instrumentation plan, which has been summarized and well documented in the *TMT Instrumentation and Performance Handbook* (see www.tmt.org), and also by Simard et al. (2010). The current plans foresee that three focal plane instruments will be available immediately when the telescope is completed. Five additional instruments are scheduled to be added during the first decade of the TMT's operation.

All three proposed first-light TMT instruments are multiobject spectrometers. One instrument (WFOS) will allow users to obtain spectra at wavelengths between 300 nm and 1,000 nm over a large FOV with spectral resolutions in the range $1,000 < R < 5,000$. An infrared imaging spectrometer (IRIS) will cover the wavelength range 0.8 μm to 2.5 μm (or possibly 0.6 μm to 5 μm). Finally, there will be an NIR MOS instrument, which, as pointed out by the authors, will essentially be a clone of the present MOSFIRE instrument of the W. M. Keck Observatory. All first-light instruments make use of the technologies of present-day large spectrometers. However, owing to the superior light-gathering power of the TMT, these instruments will surely provide important new results once the TMT gets going. In view of the technical challenges of the TMT, it is probably prudent to use relatively conventional designs for its first complement of focal-plane instruments.

The list of proposed "first-decade" instruments includes more innovative concepts. These spectrometers are expected to use MEMS-based MOS units, and most are based on the planned AO capability of the TMT. Among the interesting new ideas is a proposal by a team of the University of Colorado for the high-resolution optical spectrometer (HROS) of the TMT, which is specified to reach spectral resolutions up to $R = 10^5$. This design uses multiple IFUs to generate 0.1-arcsec-wide pseudo-slits. Instead of an echelle grating and a cross disperser (commonly used for spectroscopy at that high resolution), the University of Colorado design foresees splitting the light from the pseudo-slits into thirty-two wavelength bins by means of a hierarchical system of dichroic mirrors (called a *dichroic tree*). For each of these wavelength bins (which together cover the spectral range 310 nm to 1,100 nm), high-resolution spectra are produced by independent first-order grating spectrometers. Because of the narrow spectral range of the individual thirty-two spectrometers, these can be well optimized for their respective wavelength range. (For details see Froning et al., 2006.)

10.2.2 ESO E-ELT

A European counterpart of the TMT is the 39-m E-ELT project of the European Southern Observatory (www.eso.org/sci/facilities/eelt). The E-ELT is, at present, still in the planning stage. Again, though, much development work has already been invested in the telescope design and its potential instrumentation. Descriptions of planned E-ELT scientific instruments (based on proposals by various European groups) can be found in the *ESO Messenger* No. 140 (2010). A summary of the proposed E-ELT instruments has been presented by D'Odorico et al. (2010). Part of the potential future E-ELT instruments are, again, clones or modest extrapolations of existing devices. However, there are some proposals based on interesting novel concepts. Among those is the fiber-coupled spectrometer CODEX, which aims at radial velocity measurements of faint objects with mean errors of about 2 cm/s. This instrument will be based on a 1.7 m $\times$ 0.2 m (R4) echelle mosaic grating. Combined with the 39-m aperture of the E-ELT, this instrument is expected to discover many new Earth-like exoplanets in the habitable zones of solar-type planetary systems.

Most of the planned E-ELT spectrometers will make use of the E-ELT's adaptive optics system. Among these instruments is the proposed Mid-IR E-ELT Imager and Spectrograph (METIS), which is designed for medium- and high-resolution spectroscopy in the 3 μm to 14 μm range. For line spectroscopy, METIS is predicted to reach sensitivities comparable to those of the James Webb Space Telescope. However, owing to the 39-m telescope aperture of the E-ELT, METIS will have a much superior angular resolution.

Another proposed instrument that plans to make use of the high diffraction-limited angular resolution of the E-ELT is MICADO. This instrument will combine NIR imaging with medium resolution ($R \approx 3{,}000$) spectroscopy. According to the proposers, MICADO will greatly improve our knowledge of the spatial details of high-redshift galaxies. Moreover, MICADO will make it possible to derive accurate proper motions of faint galactic objects, which are too faint for space-based astrometric instruments.

Figure 10.1 shows an artist's concept of the two instruments, METIS and MICADO, on one of the two Nasmyth platforms of the E-ELT design. Although the spectrometers of the E-ELT will be big, their dimensions do not exceed those of the largest coudé spectrographs of the photographic era. That spectrometers for a 40-m class telescope can be built with reasonable sizes is largely due to the use of adaptive optics.

10.2.3 The James Webb Space Telescope

The most ambitious of the presently planned future space facilities is the James Webb Space Telescope (JWST) (see www.jwst.nasa.gov). According to current plans, the JWST will replace the Hubble Space Telescope by about 2018. The three main scientific instruments of the JWST will be imaging spectrometers.

Figure 10.1. Artist's concept of two proposed focal plane instruments on the Nasmyth platform of the planned E-ELT telescope. The spherical cryostat on the left encloses the MIR spectrometer METIS. In the lower right (on a circular pallet), the NIR high-resolution imager and spectrometer MICADO is visible. The larger polygonal structure above MICADO houses the proposed multi-conjugate adaptive optics relay (MAORY) system of the E-ELT. The technician in the lower image center gives scale. Image courtesy European Southern Observatory.

All will use semiconductor arrays as detectors. In addition to the three main science instruments, JWST will include a low-resolution ($150 < R < 700$) slitless NIR spectrometer, denoted NIRISS. This instrument will be part of the the telescope guiding camera system, which is provided by the Canadian Space Agency. Although the telescope is still under construction, many components of the scientific instruments do already exist.

The scientific instruments are housed in an integrated science instrument module, which is attached to the back of the main-mirror cell of the JWST (see Figure 10.2) Here the spectrometers are shielded from the solar radiation, and they can be kept cold passively by radiating into the dark space.

One of the three main instruments, the near infrared camera (NIRCam), has been designed mainly for imaging and photometry in the wavelength bands 0.6 μm to 2.3 μm and 2.4 μm to 5 μm. However, using a grism, in

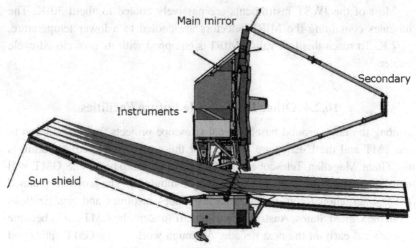

Figure 10.2. CAD image of the James Webb Space Telescope. The scientific instruments are housed in the module that is attached to the back of the main mirror cell. There the instruments are shielded from sunlight, and can be cooled passively to about 30 K. Parts of the mid-IR spectrometer are operated at an even lower temperature of ≈ 7 K, which is achieved by means of a closed-cycle cooling system. Image courtesy NASA.

the longer wavelength band low-resolution ($R \approx 2,000$) spectroscopy will be possible, too.

The near-infrared spectrograph (NIRSpec) of the JWST will be a camera and MOS spectrometer for the 0.7 μm to 5 μm range. In spectroscopic mode it will cover a field of 3×3 arcmin^2. Spectral resolutions will be $R \approx 100$ if a prism is used, or up to $R = 2,700$ with gratings. For the MOS observations there exists a choice between fixed slits, small ($3'' \times 3''$) IFUs, and MEMS-based microshutter arrays. The interesting microshutter development for NIRSpec will be discussed in Section 10.3.

The most complex JWST focal-plane instrument is the mid-infrared instrument (MIRI) for the wavelength range 5 μm to 28 μm. Among its operating modes are low-resolution ($R \approx 100$) prism spectroscopy and medium-resolution (up to about $R = 3,700$) grating spectroscopy. In grating-spectrometer mode, MIRI will use a set of four IFUs that, in each case, are equipped with three different gratings. A good description of MIRI (with instructive figures of its complex optical layout) can be found at the University of Arizona Web page of MIRI (ircamera.as.arizona.edu/MIRI/).

Most of the JWST instruments are passively cooled to about 30 K. The modules containing the MIRI detectors are cooled to a lower temperature, ≤ 7 K. To reach this low value, MIRI is equipped with its own closed-cycle cooler.

10.2.4 Other Large-Scale Future Facilities

Among the large ground-based optical telescope projects that (in addition to the TMT and the E-ELT) are planned for the near or intermediate future is the Giant Magellan Telescope (GMT) (see www.gmto.org). The GMT will consist of seven 8-m telescopes in a common mounting. The project is planned by an international consortium, which includes institutes and organizations from the United States, Australia, and Asia. If funded, the GMT could become operational early in the next decade. Although work on the GMT optics and preparatory work on the GMT site (at Las Campanas, Chile) has been started, the plans for the scientific instrumentation of the GMT are less advanced than in the cases of the TMT and the E-ELT.

A promising future X-ray spectroscopy project is the Astro-H satellite (see Takahashi et al., 2010). Astro-H is being planned by the Japan Aerospace Exploration Agency (JAXA) in partnership with U.S., Canadian, and European institutions. Currently its launch is foreseen for 2014. Like its predecessor, Suzaku (see Section 7.3), Astro-H will be equipped with photon-energy sensitive TES microcalorimeters. With lower operating temperatures, Astro-H aims at achieving an even higher spectral resolution than had been possible with the original Suzaku TES detectors. For the current status of the Astro-H project, see the Web page of JAXA (www.jaxa.jp).

The outlook for other planned spectroscopic X-ray observatories appears less certain at present. The development work for projects such as XEUS and IXO has been suspended, and the ATHENA X-ray observatory, which is considered by a working group of ESA, is still in an early planning stage.

A potentially quite interesting FIR/submillimeter space project is the SPICA observatory (see Nakagawa, 2010). With a 3.2-m main mirror, actively cooled to about 4.5 K, SPICA would have an unprecedented sensitivity in the MIR and FIR spectral range. Some development work for this project has been carried out in Japan, Europe, and the United States, but so far none of the agencies involved has made a firm commitment for this project.

Plans for new very large international facilities exist also for the radio regime (see www.skatelescope.org) and for the TeV gamma range (www.cta-observatory.org). These projects are based on well-developed and convincing science cases. However, the technical planning of these projects is not yet at the stage at which their spectroscopic potentials could be assessed reliably.

10.3 New Technologies

The long history of astronomy includes many examples in which new technologies resulted in decisive scientific breakthroughs. Thus, keeping an open mind for new technical developments is particularly important for the progress in astronomy. This section is devoted to some new technical developments that may have a significant impact on astronomical spectroscopy during the coming years or decades.

10.3.1 MEMS-Based Multiobject Spectroscopy

Many current optical spectrometers are designed for multiobject observations. As pointed out in Section 4.6.1, these MOS instruments use either laser-cut masks, individually movable slit jaws, movable IFUs, or movable fibers in a telescope focal plane. All these systems require complex mechanisms that are costly and inconvenient to operate. Some (such as slit arrangements based on exchangeable masks) are not practicable for space observatories. Therefore, during the past years, significant efforts have been devoted to developing simpler and more flexible MOS slit devices based on micro-electromechanical systems (MEMS). MOS spectrometers based on MEMS make use of either micromirrors (e.g., Canonica et al., 2010) or microshutter arrays (e.g., Li et al., 2010).

At the time of this writing, the most advanced of the MEMS-based MOS devices is the microshutter array that has been developed by NASA/GSFC for the NIRSpec instrument of the JWST. This device consists of four almost quadratic subunits with a length of about 3.6 cm. Each of these units has 365×171 trapdoor-like microshutters, which measure 0.1 mm by 0.2 mm each. Figure 10.3 shows a section of one microshutter array in the closed state. The individual microshutter lids are made of silicon nitride membranes and have torsion hinges at one of their short ends. The arrays are produced starting from silicon wafers, applying the techniques that are customarily used for manufacturing electronic circuits. The individual shutters are coated with a magnetic material; they are opened and closed by a moving magnet. They are kept in position by electrostatic forces. Under computer control, any slit pattern can be generated that is compatible with the size and the geometry

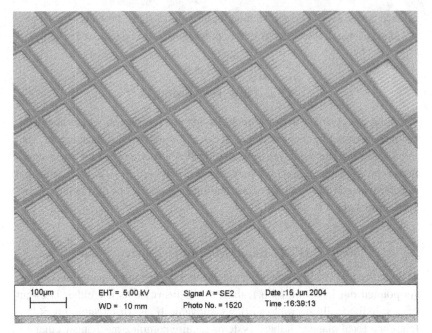

Figure 10.3. Section of the microshutter array that has been developed for the JWST NIRSpec instrument. All 100 μm × 200 μm microshutters are shown in the closed position. Image courtesy NASA.

of the individual microshutters. A detailed description of the device has been given by Li et al. (2010).

Experimental microshutter arrays have also been produced by other groups, and various industrial enterprises are working on such devices for different purposes. However, the NASA/GSFC device is, at present, the only system that has been already extensively and successfully tested in the lab, and which is ready to be used at a telescope. Although the flexibility and ease of operation of MEMS-based slit units obviously have great advantages, their future will depend on the experience at the JWST, on the costs, and on the general availability of such devices.

10.3.2 New Detectors

In the context of optical and X-ray spectroscopy, several types of detectors have been described that measure directly the energy (and thus the frequency) of individual observed photons. Such detectors can be used for low-resolution spectroscopy, or (as noted in Section 7.4), they can replace the cross dispersers in high-resolution spectrometers. As pointed out in Chapter 7, photon-energy sensitive detectors are standard devices in X-ray astronomy. However, new technical developments may soon extend their application to the whole optical range and result in a significantly better energy resolution at X-ray wavelengths.

Among the most promising of the devices that are under development at present are the microwave kinetic inductance detectors (MKIDs). These detectors are closely related to the superconducting tunnel junctions, which were described in Section 5.3. As in the case of the classical STJs, in the MKIDs Cooper pairs in a superconducting metal are dissociated by photon absorption. Because the number of unpaired electrons that are generated in this way is proportional to the quotient of $h\nu/\Delta_s$ (where Δ_s is the superconductor band gap), by measuring the total charge of the unpaired electrons the photon energy can be determined. The main difference between conventional STJs and the MKIDs is the way in which the charges are recorded. In the MKIDs one makes use of the fact that the charges modify the kinetic inductance and the microwave surface resistance of superconducting metal films. These changes are measured by irradiating the detector with a microwave beam that is in resonance with the detector. Conventional techniques for manufacturing integrated circuits can be adapted to produce arrays of MKIDs, in which the individual detector elements have different resonance frequencies. To read out such a detector array, it is exposed to a microwave beam with a drifting frequency. From the exact frequency at which a response is recorded, the output signal can be attributed to an individual detector element. In this way, large arrays with many pixels can be read out conveniently without the need of signal lines to the individual detector elements. Detailed descriptions of the MKID technology have been given, e.g., by Mazin et al. (2010) and by Bock et al. (2009). The feasibility of energy-sensitive photon detection at visual and infrared wavelengths with MKIDs has been demonstrated in the laboratory. Spectral resolutions of MKIDs are expected to reach $R \approx 400$ at a wavelength of 400 nm. First tests at a telescope are being carried out at the time of this writing (see Mazin et al., 2010). MKID-based cameras are also being developed for imaging at submillimeter wavelengths (see Schlaerth et al., 2010).

A related type of new superconducting light sensors is the quantum capacitance detector (QCD; see, e.g., Shaw et al., 2009). QCDs are essentially superconducting microbolometer arrays that use the microwave technique of the MKIDs for reading out the signal. Though promising, these devices are still in an early development stage.

A third novel type of detector with a significant potential in astronomy is arrays of miniature versions of the hot-electron bolometers (which were

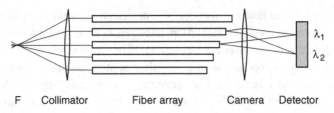

Figure 10.4. Principle of an arrayed waveguide grating spectrometer.

described briefly in Section 9.1.3). Laboratory experiments indicate that these "nano-HEBs" can extend the detection of single photons to the MIR and even FIR wavelength range (see, e.g., Karasik et al., 2010). With this technology, photon-counting FIR cameras with some energy resolution may become feasible in the future. Although the laboratory results are encouraging, the practical value of these devices has yet to be demonstrated by actual astronomical observations.

10.3.3 Arrayed Waveguide Spectrometers

Spectral devices also play an important role for communication systems using optical fibers. For these applications, small spectrometers have been developed, which can be integrated in electronic circuit boards. They are based on arrayed waveguide gratings (AWGs), and are sometimes called *integrated photonic spectrographs*, or "spectrographs on a chip." In recent years several groups have suggested adapting these devices for astronomical spectroscopy (see, e.g., Lawrence et al., 2010; Cvetojevic et al., 2012).

The principle of AWGs is indicated schematically in Figure 10.4. An AWG consists of an array of parallel single-mode waveguides, such as single-mode optical fibers. Because AWGs are manufactured using integrated-circuit technologies, they are flat devices – that is, all waveguides are placed in a single plane. Outside the AWG, the waves are retained in this plane by means of suitable flat waveguides.

As illustrated in Figure 10.4, by using arrays of waveguides of different lengths with a constant-length increment between adjacent waveguides, incremental phase shifts can be produced at the output of an AWG. By interference of the rays from the individual waveguides, an AWG can produce spectra in the same way as a grism (see Figure 3.13). However, because optical fibers and other waveguides can be curved, AWGs provide more design parameters than grisms. Moreover, the properties of the single-mode fibers can be used to ensure a plane-wave geometry at the output.

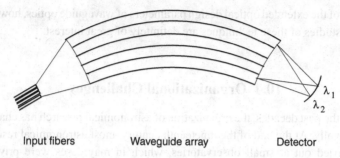

Figure 10.5. Schematic layout of a spectrometer based on a waveguide array.

In practice, the waveguides of AWGs are curved (as indicated in Figure 10.5). In this way, incremental phase shifts between adjacent waveguides can be produced while the fiber ends on both sides are aligned along straight lines or along arbitrarily curved surfaces. By arranging the fiber or waveguide ends and the detector surface along curved lines that follow the conditions of a Rowland circle (see Section 4.4.5), no extra collimator and camera optics are required, and simple spectrometers of the type indicated schematically in Figure 10.5 can be realized.

Arrayed waveguide gratings of the type outlined schematically in Figure 10.5 are widely used in optical fiber communication links for wavelength multiplexing. They are designed for NIR wavelengths, and the on-chip waveguides are based on silicon/silica (Si/SiO_2) combinations. A typical device with 428 waveguides operated at the diffraction order $m = 27$ gives a maximal theoretical spectral resolution of $R \approx 10^4$ (see Equation 3.18).

When applied for communication purposes, AWGs are used to sort signals that arrive at different wavelengths. Therefore, the input and output consist of arrays of single-mode fibers, and an output fiber array takes the place of the detector (D) indicated in Figure 10.5. Except for the unrealistic case of an ideal AO system, in astronomical applications a direct input through single-mode fibers is not practicable. Therefore, experimental astronomical AWG spectrometers have been using devices called *photonic lanterns* to feed the light of a multimode fiber from a telescope into an array of single-mode fibers, which, in turn, is used to feed the AWG (see Leon-Saval et al., 2010).

In recent years, a significant amount of laboratory work has been devoted to clarifying the applicability of modified communication AWGs for astronomical spectroscopy. However, apart from spectroscopy of the NIR night sky (which is not a typical astronomical target), no practical applications to astronomical sources have been reported so far. Therefore, at present it is still difficult to assess the potential of such waveguide-based spectral devices for astronomy.

In view of the extended optical design parameters of waveguide optics, however, further studies of these techniques are definitely of great interest.

10.4 Organizational Challenges

During the past decades, the organization of astronomical research has changed dramatically. At the end of the nineteenth century, most astronomical research was carried out at small observatories, which in may cases were privately funded and operated. Some of the technically most advanced instruments (such as the 6-foot telescope built in 1845 by the Earl of Rosse at Birr Castle, Ireland) were designed and operated by devoted amateurs. With the increasing size, costs, and complexity of modern astronomical instrumentation, however, during the twentieth century observational astronomy became more and more concentrated in fewer and larger observatories. Today, much of the progress in astronomy is based on a few very large facilities on the ground and in space. All the space instruments and most of the contemporary large ground-based observatories are funded publicly. Some of these facilities (such as the ALMA observatory) are truly global enterprises, which include the worldwide scientific communities in the respective fields. Other organizations, such as ESO or ESA, combine the joint efforts of many countries.

Whereas much of the observational research has been concentrated to fewer facilities, the number of researchers has been growing. Thus, the demand for observing time at the top facilities has been increasing enormously. The large increase in the number of observers per facility has been compensated, in part, by placing the new ground-based observatories at better sites, where more of the available time can be used productively. Moreover, the optimizing of the scheduling and the execution of observations have greatly improved the efficiency of these observatories. Another factor that helped to alleviate the pressure for observing time is that, today, observing programs are planned by larger groups, often with an international composition. Such teams can have the advantage of preparing and evaluating the observations faster, and they have the capability of establishing effective internal quality control mechanisms. Finally, an oversubscription of the large facilities has the positive effect that the programs that are expected to be most fruitful for the scientific progress can be selected for execution. Of course, this works well only if the time allocation is based strictly on scientific arguments, and if the program selection committees are composed of scientifically competent members. Membership in a time allocation committee (TAC) always involves much work and usually carries no advantages for the personal career of a scientist. Thus, it is not always easy to recruit the best astronomers for a TAC. However, for the success of a big observatory, the quality of its TAC can be as important as the quality of its instruments.

The concentration of astronomical observations toward large facilities also has had consequences for the way new instrumental developments take place. In the past, spectrometers typically were built by astronomers and designed for the special needs of their own scientific interests. At modern large facilities, however, the instruments must answer the needs of a large community of astronomers who work in different fields. Therefore, the concepts of new instruments are usually developed by committees that represent the potential users, whereas the actual construction is carried out by different teams of astronomers and engineers, often in cooperation with industry. This approach has the advantage that ideas from a large community can be included and that the industrial standards for the reliability and safety of technical devices are followed, which tends to be quite helpful for the reliable performance of complex spectrometers. However, an individual user can no longer expect that a spectrometer has the optimal properties for his or her particular application. Instead, the user must make the best of an instrument that is based on a compromise for various different tasks. To get optimal results under these conditions, it is important that the user understands the functioning, possibilities, and limitations of a spectrometer. Thus, although most contemporary observers may never be involved in building spectrometers, understanding their technologies is still an important prerequisite for the successful use of these instruments.

The problems in the operation of large modern facilities mentioned in the previous paragraphs must not be underestimated. However, for the time being, they appear to be solvable. Less clear, at present, is the issue of the accessibility of the data collected at different observatories. This concerns spectroscopy in particular. As pointed out before, spectra always contain a large amount of information, of which often only a minor part is used by the scientists who carry out the observations. Therefore, it is important that all spectroscopic data of astronomical sources are archived and are made available to other researchers. Although this is done in an exemplary way by some institutions, in other cases archives are accessible only to closed communities, or their use for outsiders is difficult. Sometimes spectral data are stored in formats that are inconvenient for further analysis or that cannot be easily converted into more useful formats. Improving this situation is an important task of the virtual observatories (VOs) that are being developed at present. However, the data access through the VOs is still very incomplete, and sometimes complex as well.

By facilitating the access to the vast amount of spectroscopic data, which by now is stored in archives, valuable observing time at the present large facilities could be saved. Thus, all observers who obtain new astronomical spectra should make sure that their data are properly archived and are made available to the interested community. This request also applies to observations carried out with smaller telescopes. Even though most of the recent progress in astronomy indeed has come from the large facilities, observations with small telescopes still have a significant impact. An example is the first spectroscopic detection of extrasolar planets, which was made with a 1.9-m telescope. Many of the institutions that operate the smaller (but, if properly equipped, still valuable) telescopes do not have efficient pipelines for archiving data. In these cases, the individual observer must take the initiative to communicate his or her results to one of the international astronomical data centers.

The technical developments of the past decades clearly have changed the way in which astronomical spectroscopy is being carried out. However, if the work at the observatories is organized according to the requirements and conditions that follow from the new realities, the new technologies provide unprecedented new opportunities for further progress in astronomy.

Appendix
List of Acronyms

Acronym	Expansion
AC	Alternating current
ACIS	Advanced CCD Imaging Spectrometer
ADC	Atmospheric dispersion compensator
ADS	Astrophysical Data System
ADU	Analog-to-digital unit
AGN	Active galactic nucleus
ALMA	Atacama Large Millimeter Array
AO	Adaptive optics
AOD	Acousto-optical deflector
AOP	Astrophysical Observatory Potsdam
AOS	Acousto-optical spectrometer
APEX	Atacama Pathfinder Experiment
ASCII	American Standard Code for Information Interchange
ATHENA	Advanced Telescope for High Energy Astrophysics
AWG	Arrayed waveguide grating
BIPM	Bureau International de Poids et Mesures
BSR	Barycentric standard of rest
CARMENES	Calar Alto High-Resolution Search for M Dwarfs with Exoearths with NIR and Optical Echelle Spectrographs
CCD	Charge-coupled device
CDS	Centre de Donée Astronomique de Strasbourg
CGRO	Compton Gamma Ray Observatory
CMB	Cosmic microwave background

CMOS	Complementary metal oxide semiconductor
CODEX	Cosmic Dynamics and Exo-Earth Experiment
COS	Cosmic Origins Spectrograph
CTS	Chirp-transform spectrometer
DC	Direct current
DROID	Distributed readout imaging device
ECDL	External cavity diode laser
E-ELT	European Extremely Large Telescope
EG	Echelon grating
EPIC	European Photon Imaging Camera
ESA	European Space Agency
ESO	European Southern Observatory
ETC	Exposure time calculator
EUV	Extreme ultraviolet
EUVE	Extreme Ultraviolet Explorer
EXES	Echelon Cross Echelle Spectrograph
FEROS	Fiber-Fed Extended Range Optical Spectrograph
FFT	Fast Fourier transform
FFTS	Fast Fourier transform spectrometer
FIFILS	Field Imaging Far-Infrared Line Spectrometer
FIR	Far infrared
FITS	Flexible Image Transport System
FOCAS	Faint Object Camera and Spectrograph
FORS	Focal Reducer Spectrograph
FOV	Field of view
FP	Fabry-Perot
FPS	Fabry-Perot spectrometer
FSR	Free spectral range
FT	Fourier transform
FTS	Fourier transform spectrometer
FUSE	Far Ultraviolet Spectroscopic Explorer
FUV	Far ultraviolet
FWHM	Full width at half maximum
GAIA	Global Astrometric Interferometer for Astrophysics
GALEX	Galaxy Evolution Explorer

GMT	Giant Magellan Telescope
GREAT	German Receiver for Astronomy at Terahertz Frequencies
GSFC	Goddard Space Flight Center
HARPS	High-Accuracy Radial Velocity Planetary Searcher
HCL	Hollow cathode lamp
HEB	Hot electron bolometer
HEMT	High-electron mobility transistor
H.E.S.S.	High Energy Stereoscopic System
HET	Hobby-Eberly Telescope
HETGS	High-Energy Transmission Grating Spectrometer
HIFI	Heterodyne instrument for the far infrared
HIRES	High-Resolution Echelle Spectrometer
HRDG	Holographically recorded diffraction grating
HRI	High-resolution X-ray imager
HROS	High-resolution optical spectrometer
HST	Hubble Space Telescope
HUT	Hopkins Ultraviolet Telescope
IAU	International Astronomical Union
IF	Intermediate frequency
IFS	Integral field spectroscopy
IFU	Integral field unit
INTEGRAL	International Gamma-Ray Astrophysics Laboratory
IPAC	Infrared Processing and Analysis Center
IR	Infrared
IRAF	Image Reduction and Analysis Facility
IRAS	Infrared Astronomical Satellite
IRIS	Infrared Imaging Spectrometer
IRS	Infrared spectrometer
IUE	International Ultraviolet Explorer
IVOA	International Virtual Observatory Alliance
JAXA	Japanese Aerospace Exploration Agency
JCMT	James Clark Maxwell Telescope
JWST	James Webb Space Telescope
KMOS	K-band Multi-Object Spectrometer

LABOCA	Large Apex Bolometer Camera
LAMOST	Large Area Multi-Object Fiber Spectroscopic Telescope
LAT	Large Area Telescope
LBL	Lawrence Berkeley National Laboratory
LBT	Large Binocular Telescope
LETGS	Low-Energy Transmission Grating Spectrometer
LO	Local oscillator
LOS	Line of sight
LRIS	Low-Resolution Imaging Spectrograph
LSR	Local standard of rest
LUCIFER	LBT Near Infrared Utility with Camera and Integral Field Unit for Extragalactic Research
MCP	Microchannel plate
MEMS	Micro-Electromechanical system
METIS	Mid-IR E-ELT Imager and Spectrograph
MICADO	Multi-adaptive Optics Imaging Camera for Deep Observations
MICHELLE	Mid-Infrared Echelle Spectrometer
MIDAS	Munich Image Data Analysis System
MIRI	Mid-infrared instrument
MK(K)	Morgan-Keenan-Kellman (spectral classification)
MKID	Microwave kinetic induction detector
MODS	Multiobject double spectrograph
MOIRCS	Multi-Object Infrared Camera and Spectrometer
MOS	Multiobject spectroscopy
MOSFIRE	Multi-Object Spectrometer for Infrared Exploration
NASA	National Aeronautic and Space Administration
NIRCam	Near Infrared Camera
NIRISS	Near Infrared Imager and Slitless Spectrograph
NIRSpec	Near Infrared Spectrograph
NIST	National Institute of Standards and Technology
NOAO	National Optical Astronomical Observatories
NRL	Naval Research Laboratory
NUV	Near ultraviolet
OAO	Orbiting astronomical observatory
OB	Observing block

ORFEUS	Orbiting Retrievable Far and Extreme Ultraviolet Satellite
OSO	Orbiting solar observatory
PACS	Photoconducting Array Camera and Spectrometer
PEPSIOS	Polyetalon pressure scanned interferometric optical spectrometer
PT	Piezoelectric transducer
QCD	Quantum capacitance detector
QE	Quantum efficiency
QSO	Quasi-stellar object
RF	Radiofrequency
SAFIRE	Submillimeter and Far-Infrared Experiment
SCUBA	Submillimeter Common User Bolometer Array
SDSS	Sloan Digital Sky Survey
SI	Système internationale d'unites
SIS	Superconducting-insulator-superconducting
SKA	Square kilometer array
S/N	Signal-to-noise ratio
SOFIA	Stratospheric Observatory for Infrared Astronomy
SOHO	Solar and Heliospheric Observatory
SPICA	Space Infrared Telescope for Cosmology and Astrophysics
SPIE	Society of Photo-Optical Instrumentation Engineers
SPIFFI	Spectrometer for Infrared Faint Field Imaging
SPIRE	Spectral and Photometric Imaging Receiver
SQUID	Superconducting quantum interference device
STIS	Space Telescope Imaging Spectrograph
STJ	Superconducting tunnel junction
TAC	Time allocation committee
TES	Transition edge sensor
THIS	Tunable heterodyne infrared spectrometer
TMT	Thirty-meter Telescope
UV	Ultraviolet
UVES	UV-Visual Echelle Spectrograph

VIMOS	Visible Multiobject Spectrograph
VIRUS	Visible Integral Field Replicable Unit Spectrograph
VLT	Very Large Telescope
VO	Virtual observatory
VPG	Volume phase grating
VPHG	Volume phase holographic grating
WD	White dwarf star
WFOS	Wide-Field Optical Spectrometer
XMM	X-ray Multimirror Mission

References

Appenzeller, I. 1989. Detectors and Receivers. Pages 299–350 of: Appenzeller, I., Habing, H. J., and Lená, P. (eds.), *Evolution of Galaxies, Astronomical Observations*. Springer Lecture Notes in Physics, Vol. 333.

Appenzeller, I. 2009. *High-Redshift Galaxies*. Springer.

Appenzeller, I., et al. 1998. *The ESO Messenger*, **94**, 1.

Araujo-Hauck, C., et al. 2007. *The ESO Messenger*, **129**, 25.

Arnaud, K., Smith, R., and Siemiginowska, A. 2011. *Handbook of X-Ray Astronomy*. Cambridge University Press.

Atwood, W. B., et al. 2009. *ApJ*, **697**, 1071.

Avila, G. and Singh, P. 2008. *SPIE*, **7018**, 157.

Bacon, R., et al. 1995. *A & AS*, **113**, 347.

Bahner, K. and Solf, J. 1972. Design Study of the Coudé Spectrographs for the 2.2-m Telescopes of the MPI for Astronomy. In: Laustsen, S. and Reiz, A. (eds.), *Proc. ESO/CERN Conference on Auxiliary Instrumentation for Large Telescopes* (ESO Conference Proceedings), p. 247.

Baranne, A. and Duchesne, M. 1972. Montage à pupille blanche pour caméra électronique. In: Laustsen, S. and Reiz, A. (eds.), *Proc. ESO/CERN Conference on Auxiliary Instrumentation for Large Telescopes* (ESO Conference Proceedings), p. 241.

Barden, S. C., Arns, J. A., and Colburn, W. S. 2000. *PASP*, **112**, 809.

Barnes, S. I. and MacQueen, P. J. 2010. *SPIE*, **7735**, 204.

Barnstedt, J., et al. 1999. *A & AS*, **134**, 561.

Bass, M., ed. 1995. *Handbook of Optics*, 2nd ed., Vols. 1–4. McGraw-Hill.

Baudry, A. 2011. *ALMA Newsletter*, **7**, 18.

Bazarghan, M. and Gupta, R. 2008. *Astrophys. & Space Sciences*, **315**, 201.

Bean, J. L., et al. 2010. *ApJ*, **713**, 410.

Becker, W. 2009. X-Ray Emission from Pulsars and Neutron Stars. Chapter 6 of: Becker, W. (ed.), *Neutron Stars and Pulsars*. Springer.

Beletic, J. E., Beletic, J. W., and Amico, P., eds. 2006. *Scientific Detectors for Astronomy 2005*. Springer.

Benford, D. J., et al. 2003. *SPIE*, **4857**, 105.

Bintley, D., et al. 2010. *SPIE*, **7741**, 2.

Bland-Hawthorn, J., et al. 2010. *SPIE*, **7735**, 22.

Bock, J. J., et al. 2009. *Superconduction Detector Arrays for Far-Infrared and mm-Wave Astrophysics.* The Astronomy and Astrophysics Decadal Report. Technology Development Paper No. 45 (see ADS).

Born, M. and Wolf, E. 1987. *Principles of Optics.* Pergamon Press.

Bowen, I. S. 1938. *ApJ*, **88**, 113.

Bradford, C. M., et al. 2010. *SPIE*, **7731**, 2.

Brandt, W. N. and Hasinger, G. 2005. *An. Rev. Astr. Astrophys.*, **43**, 827.

Burney, J. A. 2007. *Transition-edge sensor imaging arrays for astrophysical applications.* PhD thesis, Stanford University, Palo Alto, CA.

Burnight, T. R., et al. 1949. *Phys. Rev.*, **76**, 165.

Campbell, R. D. and Thompson, D. J. 2006. *The Effects of Charge Persistence in Aladdine III InSb Detectors on Scientific Observations.* In: Beletic, J. E., Beletic, J. W., and Amico, P., eds., *Scientific Detectors for Astronomy 2005.* Springer.

Canonica, M., et al. 2010. *SPIE*, **7739**, 151.

Carleton, N., ed. 1976. *Methods of Experimental Physics*, **12**, Part 1, Chapters 10–12. Academic Press.

Clénet, Y., et al. 2002. *PASP*, **114**, 563.

Colpi, M., et al., eds. 2009. *Physics of Relativistic Objects in Compact Binaries.* Springer.

Cottam, J., Paerels, F., and Mendez, M. 2002. *Nature*, **420**, 51.

Cox, A. N. 2000. *Allen's Astrophysical Quantities.* Springer.

Craig, N., et al. 1997. *ApJS*, **113**, 131.

Crowther, P. A. 2007. *An. Rev. Astr. Astrophys.*, **45**, 177.

Cui, X., et al. 2008. *SPIE*, **7012**, 3.

Cvetojevic, N., et al. 2012.*Optics Express*, **20**, 2062.

de Graauw, Th., et al. 2010. *A & A*, **518**, L6.

Dekker, H., D'Odorico, S., and Fontana, A. 1994. *The ESO Messenger*, **76**, 16.

Dekker, H., et al. 2000. *SPIE*, **4008**, 534.

Dekker, H., et al. 2003. *SPIE*, **4842**, 139.

D'Odorico, S., et al. 2010. *The ESO Messenger*, **140**, 17.

Dopita, M. A. and Sutherland, R. S. 2003. *Astrophysics of the Diffuse Universe.* Springer.

Eisenhauer, F., et al. 2003. *The ESO Messenger*, **113**, 17.

Euchner, F., et al. 2006. *A& A*, **451**, 671.

Ewen, H. I. and Purcell, E. M. 1951. *Nature*, **168**, 256.

Falgarone, E., et al. 2008. *A& A*, **487**, 247.

Fellgett, P. B. 1949. *J. Opt. Soc. Am.*, **39**, 970.

Frank, J., King, A., and Raine, D. 2003. *Accretion Power in Astrophysics*, 3rd ed. Cambridge University Press.

Frank, S., Appenzeller, I., Noll, S., and Stahl, O. 2003. *A & A*, **407**, 473.

Fraunhofer, J. 1817. *Denkschriften der Münchener Akademie der Wissenschaften*, **5**, 193.

Friedman, H., Byram, E. T., and Chubb, T. A. 1966. *Science*, **153**, 1527.

Friedman, H., Lichtman, S. W., and Byram, E. T. 1951. *Phys. Rev.*, **83**, 1025.

Froning, C., et al. 2006. *SPIE*, **6269**, 61.

Giacconi, R. 2003. *Rev. Modern Physics*, **75**, 995.

Giacconi, R., Gursky, H., Paolini, F., and Rossi, B. B. 1962. *Phys. Rev. Lett.*, **9**, 439.

Glass, I. S. 1999. *Handbook of Infrared Astronomy.* Cambridge University Press.

Glasse, A. C., Atad, E., and Montgomery, D. 1993. *ASPC*, **41**, 401.

Gom, B. and Naylor, D. 2010. *SPIE*, **7741**, 67.

Grant, E. 1974. *A Source Book of Medieval Science*. Harvard University Press, p. 864.

Gray, R. O. and Corbally, C. J. 2009. *Stellar Spectral Classification*. Princeton University Press.

Grewing, M., et al. 1991. *The ORFEUS Mission*. In: Malina, R. F. and Bowyer, S., eds. *Extreme Ultraviolet Astronomy*. Pergamon Press, p. 437.

Griffin, M., et al. 2010. *A & A*, **518**, L3.

Hamuy M., et al. 1992. *PASP*, **104**, 533.

Hamuy, M., et al. 1994. *PASP*, **106**, 566.

Hartigan, P., Edwards, S., and Pierson, R. 2004. *ApJ*, **609**, 261.

Hatzes, A. and Cochran, W. 1992. In: Ulrich, M. (ed.), *Proc. ESO Workshop on High-resolution Spectroscopy with the VLT*. ESO Conference Proceedings Vol. **40**, 275.

Hearnshaw, J. B. 1986. *The Analysis of Starlight*. Cambridge University Press.

Hearnshaw, J. B. 2009. *Astronomical Spectrographs and Their History*. Cambridge University Press.

Hernandez, O., et al. 2008. *PASP*, **120**, 665.

Herrera-Martinez, G., et al. 2009. *Revista Mex. AA* (Serie de Conferencias), **37**, 156.

Heyminck, S., et al. 2008. *SPIE*, **7014**, 33.

Hill, G. J. and MacQueen, I. J. 2002. *SPIE*, **4836**, 306.

Hill, G. J., Wolf, M. J., Tufts, J. R., and Smith, E. C. 2003. *SPIE*, **4842**, 1.

Hiltner, W. A. 1964. *Astronomical Techniques*. Chicago University Press.

Hinton, J. A. and Hofmann, W. 2009. *An. Rev. A&A*, **47**, 523.

Homes, C. C., et al. 2007. *Applied Optics*, **46**, 7884.

Horne, K. 1986. *PASP*, **98**, 609.

Houck, J. R., et al. 2004. *ApJS*, **154**, 18.

Howell, S. B. 2006. *Handbook of CCD Astronomy*, 2nd ed. Cambridge University Press.

Ichikawa, T., et al. 2006. *SPIE*, **6269**, 38.

Iwert, O. and Delabre, B. 2010. *SPIE*, **7742**, 60.

Jack, D., Hauschildt, P. H., and Baron, E. 2009. *A & A*, **502**, 1043.

Johns-Krull, C. M. 2007. *Proc. IAU Symposium*, **243**, 31.

Justtanont, K., et al. 2012. *A & A*, **537**, 144.

Karasik, B. S., et al. 2010. *SPIE*, **7741**, 34.

Kashikawa, N., et al. 2002. *PASJ*, **54**, 819.

Kaufer, A. 1998. *ASP Conf. Ser.*, **152**, 337.

Kaufer, A. and Pasquini, L. 1998. *SPIE*, **3355**, 844.

Kaufer, A., et al. 1999. *The ESO Messenger*, **95**, 8.

Kawada, M., et al. 2008. *PASJ*, **60**, 389.

Kayser, H. 1900. *Handbuch der Spektroskopie*. Verlag S. Hirzel.

Kirkpatrick, J. D. 2005. *An. Rev. Astr. Astrophys.*, **43**, 195.

Kitchin, C. R. 2008. *Astronomical Techniques*. CRC Press.

Klein, B., et al. 2006a. *SPIE*, **6275**, 33.

Klein, B., et al. 2006b. *A & A*, **454**, L29.

Klein, R., et al. 2010. *SPIE*, **7735**, 6.

Koenigsberger, G., et al. 2011. *IAU Symp.*, **272**, 511.

Köppen, J. 2010. *Sterne und Weltraum*, **4/2010**, 88.

Kraushaar, W. L., et al. 1965. *ApJ*, **141**, 845.

Krautter, J., Appenzeller, I., and Jankovics, I. 1990. *A & A*, **236**, 416.

Kurucz, R. L. 1970. *SAO Special Report*, **309**, 1.

Lawrence, J., et al. 2010. *SPIE*, **7739**, 144.

Le Fèvre, O., et al. 2002. *The ESO Messenger*, **109**, 21.

Léna, P, Lebrun, F., and Mignard, F. 1998. *Observational Astrophysics*. Springer.

Leon-Saval, S. G., Argyros, A., and Bland-Hawthorn, J. 2010. *Optics Express*, **18**, 8430.

Lequeux, J. 2005. *The Interstellar Medium*. Springer.

Lester, J. B. and Neilson, H. R. 2008. *A & A*, **491**, 633.

Li, M. J., et al. 2010. *SPIE*, **7594**, 20.

Littrow, O. 1863. *Berichte der Sternwarte Wien*, **47**, 26.

Longair, M. S. 2008. *Galaxy Formation*, 2nd ed. Springer Verlag, Chapters 5–8.

Mack, J. E., et al. 1963. *Applied Optics*, **2**, 873.

Malacara, D. and Thompson, B. J., eds. 2001. *Handbook of Optical Engineering*. Marcel Dekker.

Martin, D. D. E. and Verhoeve, P. 2010. *ISSIR*, **9**, 441.

Mayor, M., et al. 2003. *The ESO Messenger*, **114**, 20.

Mazin, B. A., et al. 2010. *SPIE*, **7735**, 42.

McLean, I. S., et al. 2010. *SPIE*, **7735**, 12.

Meeks, M. L., ed. 1976. *Methods of Experimental Physics*, **12**, Part 3 (Radio Observations). Academic Press.

Menten, K. M., et al. 2008. *A & A*, **477**, 185.

Morgan, W. W., Keenan, P. C., and Kellman, E. 1943. *An Atlas of Stellar Spectra, with an Outline of a Spectral Classification*. University of Chicago Press.

Morgan, W. W. and Mayall, N. U. 1957. *Publ. Astr. Soc. Pacific*, **69**, 291.

Morgan, W. W., Abt, H. A. P., and Tapscot, J. W. 1978. *Revised MK Atlas for Stars Earlier than the Sun*. University of Chicago and Kitt Peak National Observatory.

Morton, D. C. 1991. *ApJS*, **77**, 119.

Nakagawa, T. 2010. *SPIE*, **7731**, 18.

Naylor, D. A., et al. 2010. *SPIE*, **7731**, 29.

Ness, J.-U., et al. 2002. *A & A*, **394**, 911.

Newton, I. 1740. *Optice*. Bousquet et Sociorum, Lausanne/Geneva.

Noll, S. 2002. *The FORS Deep Field Spectroscopic Survey*. Ph.D. Thesis, University of Heidelberg, Germany.

Oke, J. B. 1990. *AJ*, **99**, 1621.

Oke, J. B., et al. 1995. *PASP*, **107**, 375.

Osmer, P., et al. 2000. *SPIE*, **4008**, 40.

Osterbrock, D. E., et al. 1996. *PASP*, **108**, 277.

Osterbrock, D. E., et al. 1997. *PASP*, **109**, 614.

Paerels, F. B. S. and Kahn, S. M. 2003. *An. Rev. Astr. Astrophys.*, **41**, 291.

Palmer, Ch. and Loewen, E. 2005. *Diffraction Grating Handbook*, 6th ed. Newport, Irvine, CA.

Pasquali, A., et al. 2006. *PASP*, **118**, 270.

Pepe, M., et al. 2004. *A & A*, **423**, 385.

Perruchot, S., et al. 2008. *SPIE*, **7014**, 17.

Poglitsch, A., et al. 2008. *SPIE*, **7010**, 3.

Pottschmidt, K., et al. 2012. *A Suzaku View of Cyclotron Line Sources and Candidates*. *AIPC*, **1427**, 60.

Pradhan, A. K. and Nahar, S. N. 2011. *Atomic Astrophysics and Spectroscopy.* Cambridge University Press.

Pravdo, S. H., et al. 1976. *ApJ*, **206**, L41.

Quirrenbach, A., et al. 2010. *SPIE*, **7735**, 37.

Richter, M. J., et al. 2003. *SPIE*, **4857**, 37.

Roddier, F. 1999. *Adaptive Optics in Astronomy.* Cambridge University Press.

Romani, R. W., et al. 2003. *ASPC*, **291**, 339.

Rowland, H. A. 1882. *The Observatory*, **5**, 224.

Ruder, H., Wunner, G., Herold, H., and Geyer, F. 1994. *Atoms in Strong Magnetic Fields.* Springer.

Sanchez Almeida, et al. 2010. *ApJ*, **714**, 487.

Schieder, R. T., et al. 2003. *SPIE*, **4855**, 290.

Schlaerth, J. A., et al. 2010. *SPIE*, **7741**, 4.

Seifert, W., et al. 2003. *SPIE*, **4841**, 962.

Seifert, W., et al. 2010. *SPIE*, **7735**, 256.

Sharples, R., et al. 2010. *The ESO Messenger*, **139**, 24.

Shaw, M. D., et al. 2009. *Phys. Rev. B*, **79**, 144511.

Simard, L., et al. 2010. *SPIE*, **7735**, 70.

Sion, M., et al., 1983. *ApJ*, **269**, 253.

Siringo, G., et al. 2009. *A & A*, **497**, 945.

Sonnabend, G., Wirtz, D., Vetterle, V., and Schieder, R. 2005. *A & A*, **435**, 1181.

Stacey, G. J., et al. 2004. *SPIE*, **5498**, 323.

Stahl, O., Kaufer, A., and Tubbesing, S. 1999. *ASPC*, **188**, 331.

Stetson, P. B. 1989. *Highlights of Astronomy*, **8**, 635.

Stumpff, P. 1980. *A & AS*, **41**, 1.

Takahashi, T., et al. 2010. *SPIE*, **7732**, 2.

Tanaka, Y., et al. 1994. *PASJ*, **46**, L37.

Tennyson, J. 2011. *Astronomical Spectroscopy: An Introduction to Atomic and Molecular Physics of Astronomical Spectra*, 2nd ed. World Scientific Publ. Co.

Tong, C.-Y. E., et al. 2006. *SPIE*, **6373**, 19.

Trümper, J., et al. 1978. *ApJ*, **219**, L105.

van de Hulst, H. C., Muller, C. A., and Oort, J. H. 1954. *Bull. Astron. Inst. Netherlands*, **12**, 117.

Verhoeve, P., Martin, D. D. E., and Venn, R. 2010. *SPIE*, **7742**, 16.

Villanueva, G. and Hartogh, P. 2004. *Experimental Astronomy*, **18**, 77.

Vogt, S. S., et al. 1994. *SPIE*, **2198**, 362.

Völk, H. J. and Bernlöhr, K. 2009. *Experimental Astronomy*, **25**, 173.

Walborn, N. R., et al. 2002. *Astr.J*, **123**, 2754.

Walraven, Th. and Walraven, J. H. 1972. Some features of the Leiden radial velocity instrument. In: Laustsen, S. and Reiz, A. (eds.), *Proc. ESO/CERN Conference on auxiliary instrumentation for large telescopes*, ESO Conference Proceedings, p. 175.

Williams, D. 1976. *Methods of Experimental Physics*, **13**, Part A.

Wilson, T. L., Rohlfs, K., and Hüttemeister, S. 2009. *Tools of Radio Astronomy*, 5th ed. Springer.

Wishnow, E. H., et al. 2010. *SPIE*, **7734**, 8.

York, D. G., et al. 2000. *Astr.J*, **120**, 1579.

Index